CHEMISTRY AND PHYSICS

OF CARBON

Volume 8

CHEMISTRY AND PHYSICS OF CARBON

A SERIES OF ADVANCES

Volume 8

Edited by

P. L. Walker, Jr. and Peter A. Thrower

DEPARTMENT OF MATERIAL SCIENCES
THE PENNSYLVANIA STATE UNIVERSITY
UNIVERSITY PARK, PENNSYLVANIA

1973

MARCEL DEKKER, INC.   New York

PREFACE TO VOLUME 8

The publication of Volume 8 of the <u>Chemistry and Physics of Carbon</u> series bears further testimony to the rapidly expanding science of carbon materials. It has long been felt that to ensure the maximum value of the review articles published in these volumes there must be the minimum delay between the author completing the manuscript and the final publication. In order to help meet this need this is the first volume in this series to appear under co-editorship, P. A. Thrower joining P. L. Walker, Jr. in the task of editing and preparing the manuscripts for publication. It is hoped that this step will further expedite the publication of future volumes and thus ensure their immediate relevance to the needs of research workers in the field. With this volume regular readers of this series will notice some changes in production format. These changes have been necessitated by continually rising costs and the desire to make these volumes available to a wider readership. A consequence of these changes will be the appearance of Volumes 9-11 within the next few months.

An area of theoretical interest which has received much help in recent years from the availability of well-oriented pyrolytic graphite is that of electronic properties. In the first chapter of this volume, I. L. Spain considers this important topic, the understanding of which has, to quote the author, "been a challenge to scientists since at least the beginning of this century." It was not until late in 1968, however, that direct experimental evidence was obtained suggesting that the parameter γ_2 appearing in the band theory of graphite was negative. The significances of this determination are fully discussed by the author, who also gives a comprehensive account of the experimental techniques that have been used to unravel this extremely complex problem.

The importance of pyrolytic graphites as tools for research into the basic properties of graphite is much evident from a reading of this first chapter. Indeed Dr. Spain has seen fit to limit his discussion to such materials, whose properties approach those of single crystals. A chapter in Volume 5 of this series by J. C. Bokros gave an introduction to the whole spectrum of pyrolytic materials, and it is hoped to include more information on their structure and uses in future volumes. Indeed it is planned to include contributions in Volume 9 on the rising importance of pyrolytic carbons as biomaterials, and in Volume 11 on the structure of well-oriented pyrolytic graphites.

Following a chapter on what may be considered a theoretically biased topic we are returned to the practical applications of graphite in the last two chapters. The final chapter of this volume concerns carbon fibers. It is surely not necessary to point out to readers of this series the great importance that carbon fibers have attained during the past decade. Fiber strengthened composite materials are certainly not new to our experience, but the recent availability of carbon fibers with high specific modulus and strength has opened exciting possibilities in many engineering areas, especially the aerospace industry. One of the most important problems in the manufacture of a successful composite material is the formation of a strong bond between the fiber and the matrix. This in turn is largely dependent on the surface properties of the fiber. In the second chapter of this volume, D.W. McKee and V.J. Mimeault review the methods that have been used to both monitor and modify the surface of these materials. They conclude that "much fundamental research remains to be done," and thus we anticipate that there will be further chapters on this extremely important subject in the not too distant future. Indeed it is planned to include a contribution on carbon fibers from rayon precursors in Volume 9 and one on the structure and physical properties in Volume 11.

A use of graphite which has been of major importance for a number of years is in the nuclear power industry. The earliest reactors had graphite moderators and some of the later more advanced designs used graphite in the fuel elements themselves. Here again pyrolytic carbon has been of immense practical value, as discussed in Volume 5 by Bokros. One of the major problems facing the reactor designer, with the introduction of carbon and graphite as fuel containers, was the retention of the fission products from the uranium fuel. Many of these elements produced during the fission reaction were expected to diffuse rapidly through the graphite structure and thus mix with the coolant gases. Some of these phenomena and the techniques used to investigate them are discussed by S. Yajima in the final chapter.

Peter A. Thrower
P.L. Walker, Jr.

CONTRIBUTORS TO VOLUME 8

D. W. McKee, General Electric Research and Development Center, Schenectady, New York

V. J. Mimeault, General Electric Research and Development Center, Schenectady, New York

I. L. Spain, Department of Chemical Engineering and Engineering Materials Group, University of Maryland, College Park, Maryland

Seishi Yajima, Research Institute for Iron, Steel, and Other Metals, Tohoku University, Sendai, Japan

CONTENTS OF VOLUME 8

CONTENTS OF OTHER VOLUMES

CHEMISTRY AND PHYSICS

OF CARBON

Volume 8

THE ELECTRONIC PROPERTIES OF GRAPHITE

I. L. Spain

Department of Chemical Engineering
and Engineering Materials Group
University of Maryland
College Park, Maryland

I. INTRODUCTION

An understanding of the electronic properties of graphite has been a
challenge to scientists since at least the beginning of this century. Since
the pioneering work of Roberts in 1913 [1] and Washburn in 1915 [2], a
steadily increasing number of papers has been published. Two main developments have occurred since then. First, detailed theories have been
developed of the dynamical properties of electrons in solids. Such theories
were first applied to graphite by Coulson [3] and Wallace [4], until, at
present, to quote Williamson, Foner, and Dresselhaus [5], "graphite is
the semimetal for which the band structure has probably been analyzed
most extensively on the basis of general theories--the experimental data
are used not only for establishment of dispersion relations and the Fermi
surface but also for the determination of its band parameters." Second, as
theories have developed, specialized experimental techniques have been

devised to test them. Such experiments include the De Haas-Van Alphen effect, cyclotron resonance, and magnetoreflection.

It is now well established that graphite is a semimetal, in which the highest filled valence band overlaps the lowest empty conduction band by about 36 meV. Even at temperatures approaching absolute zero there are empty valence states (holes) and filled conduction states (electrons). As a result of the large spacing between the carbon layers in the crystal structure, mechanical, thermal, and electrical properties are highly anisotropic [6]. Ratios for the principal conductivities have been reported to be as high as 8×10^4 [7,8]. It is possible that a collective description of the electrons in graphite is required to explain this high anisotropy as well as other properties. The Slonczewski-Weiss model is based on the one-electron approximation, but a start is being made on the assessment of collective effects [9,10].

In the last few years several reviews of the electronic properties have appeared. The review by Haering and Mrozowski [11] discusses the band theory of Slonczewski and Weiss [12] in detail. Since this theory is still used today, I will not duplicate this groundwork, but will present the results of the theory insofar as they are applicable to the rest of the chapter. The book by Reynolds [13] gives a general account of the properties of graphite, with a chapter on electronic properties. Attention is also drawn to a text in French, Les Carbones [14], and to other chapters [15, 16] in the present series which inevitably overlap to a small extent with the present text. McClure has given a short review of the electronic properties of graphite at the last two conferences on the physics of semimetals and narrow-band semiconductors [17, 18]. The later paper appeared in preprint form during the writing of this chapter. The review by Saunders [19] covers the transport properties of graphites and carbons with references up to 1967, whereas this chapter is restricted to highly crystallized materials with properties approaching those of ideal or single crystals.

The later paper by McClure [18] is mainly devoted to the recent controversy as to the sign of the parameter γ_2 appearing in the band theory of Slonczewski and Weiss [12], a parameter that largely controls the overlapping of the bands. The sign of this parameter determines whether the average mass of the holes is greater than that of the electrons (positive γ_2) or vice versa (negative γ_2). For reasons discussed later in this chapter a positive value had been used prior to late in 1968. It had been pointed out by several workers that galvanomagnetic data could best be described with a heavy-electron model (negative γ_2) [17]. However, Schroeder, Dresselhaus, and Javan [20], presenting new magnetoreflection data coupled with an analysis of De Haas-Van Alphen type of periods and a re-examination of cyclotron-resonance data, made a strong case for a negative value. Using a direct experimental determination, Woollam [21] has confirmed the negative value. Other support for a negative value comes from an analysis of the effect of light neutron damage on De Haas-Van Alphen periods by a group at Imperial College, London [22], and an examination of galvanomagnetic and thermo-magnetic data [23, 24]. Kechin [25], developing a theory for galvanomag-netic effects, independently concluded that γ_2 was negative. Support for the location of electrons at point K also comes from recent photoemission and secondary-electron-emission measurements [26].

In the face of such strong evidence a negative value for γ_2 is chosen throughout this chapter. This presents a potentially confusing situation, since all papers between about 1960 and late 1968 used a positive value. It may be assumed that a positive value was used in the literature published before 1968, and repetitive statements to this effect will be avoided.

Apart from being restricted to the properties of highly crystalline graphite, this chapter also concentrates on the properties of electrons with energies close to the electrochemical potential, thus largely ignoring prop-erties explored by using photons of energy greater than about 0.1 eV. This has been done in the belief that a separate review article would best cover this material.

II. CRYSTAL STRUCTURE AND THE BRILLOUIN ZONE

Several structures have been proposed for graphite, including high-pressure phases [27, 28], but it is assumed that for the crystals to be discussed here most of the solid has the Bernal structure [29]. A change in the stack-ing sequence ABAB... may appreciably affect the band structure. Calcula-tions have been reported for the ABCABC... stacking sequence [30, 31] corresponding to rhombohedral graphite. The bands have very different properties than those of the Bernal structure.

The unit cell of the structure contains four atoms (labeled ABA'B' in Fig. 1), two each in adjacent layer planes. The principal lattice constants

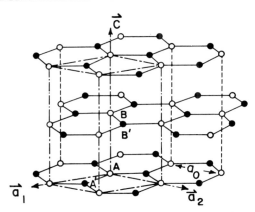

FIG. 1. The crystal structure of graphite, showing the stacking sequence ABAB.... Atoms of type α with neighbors directly above and below in adjacent planes are shown with open circles; those of type β, with full circles. The unit cell with four atoms A, B, A', B' is shown. The cell height C_0 is 6.708 Å and the basal dimension a_0 = 2.46 Å. The in-plane bond length AA' is 1.46 Å.

are a_0 = 2.46 Å, C_0 = 6.708 Å [32], so that the unit cell volume is $\sqrt{3}a_0^2C_0/2$. It is then possible to construct the Brillouin zone in two ways [11]. The symmetrical zone shown in Fig. 2 has a vertical zone edge of height $2\pi/C_0$, distance $4\pi/3a_0$ from center to zone edge, and volume $16\pi^3/\sqrt{3}a_0^2 C_0$. The zone may thus accommodate one electron of a given spin direction for every four carbon atoms. The electronic states of interest lie near the vertical zone edges HKH', and since there is only approximately one in 10^4 free electrons, the regions of reciprocal space of interest extend only about 1% of the distance from these edges toward the center of the zone.

The wavefunctions for electrons in crystalline solids are of the Bloch form

$$\psi_{n,\underline{k}}(\underline{r}) = u_{n,\underline{k}}(\underline{r})\, e^{i\underline{k}\cdot\underline{r}}, \tag{1}$$

where $\underline{k}$ is the label (wavevector) for the state in the band n, and $u_{\underline{k}}(\underline{r})$ is a periodic function such that $u_{\underline{k}}(\underline{r}) = u_{\underline{k}}(\underline{r} + \underline{T})$, where $\underline{T}$ is a principal lattice-translation vector. It follows that the label $\underline{k}$ is undefined to the extent of any principal reciprocal lattice vector $\underline{G}$ $[\psi_{\underline{k}}(\underline{r}) \equiv \psi_{\underline{k}+\underline{G}}(\underline{r})]$. It is convenient to restrict solutions to a given Brillouin zone, in which the faces have

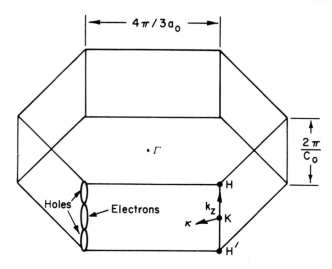

FIG. 2. The symmetrical Brillouin zone of graphite, with location of electron and hole Fermi surfaces shown. The angle a is made by the basal wavevector component $\kappa = (k_x^2 + k_y^2)^{1/2}$ with the line joining the point K with the center of the zone, Γ. Distance ΓK is $4\pi/3a_0$, and the zone height $2\pi/C_0$.

the important property that the energy eigenvalues for the electron states are discontinuous at the boundaries. However, in Fig. 2 energy surfaces are shown as "solid" figures about the vertical zone edge, so that states lie outside the first zone (extended-zone scheme). For purposes of counting states, six surfaces in the first Brillouin zone are then equivalent to two solid surfaces. The surfaces also protrude across the hexagonal faces, consistent with currently accepted values for the overlap parameters. The top of the surfaces extending into the next zone may be associated with minority carriers.

III. BAND THEORY OF GRAPHITE IN ZERO MAGNETIC FIELD

A. General Considerations

It is convenient to imagine that the electrons around the carbon nuclei are arranged into two bond types, that is, s^2p^2 hybridized orbitals (σ bonds) in the x-y plane, and p_z orbitals (π bonds) in the z direction. The covalent bonds linking atoms together in the basal planes are about two orders of magnitude stronger than the van der Waals bonds between adjacent layers. This condition leads to anisotropies in mechanical and thermal properties. It is well established both experimentally and theoretically that the valence band of graphite arises from the bonding π orbitals and the conduction band from the antibonding (π^*) orbitals. These states overlap in energy by about 30 to 40 meV, whereas the nearest σ-bond states lie several electron volts away from the region of overlap.

A detailed discussion of earlier calculations of the energy-band structure [3, 4, 12, 33-37] has been given in the review article by Haering and Mrozowski [11]. The most general treatment has been given by Slonczewski and Weiss [12], who considered in detail the general features of the bands introduced by the lattice symmetry. They concentrated only on the energy states near the vertical edges of the Brillouin zone and found the variation of energy along the zone edge by a Fourier expansion method (equivalent to a tight-binding calculation for a one-dimensional crystal). The variation with wavevector component in the basal plane $(\kappa^2 = k_x{}^2 + k_y{}^2)$ was found by using the $\underline{k} \cdot \underline{p}$ perturbation treatment. This model expresses the electron-dispersion relationship in terms of seven parameters $(\gamma_0, \gamma_1, \gamma_2, \gamma_3, \gamma_4, \gamma_5, \Delta)$ that may be related to tight-binding parameters (Table 1). It is to be noted that the out-of-plane overlap integrals are smaller in magnitude than the in-plane integral γ_0 and are modified by the difference between lattice sites of type α (with lattice sites on neighboring planes immediately above and below original site) and those of type β (without such neighbors) (Fig. 1). Calculated values of the parameters are sensitive to the crystal potential used, and since the details of the bands are very sensitive to the parameters themselves, attention has focused on the experimental determination of the parameters (see, however, [38] and [39])

In the limit that the out-of-plane parameters are reduced to zero, the results reduce to the two-dimensional case in which the crystal symmetry imposes the condition that the two π bands be degenerate at the six corners of the plane Brillouin zone [3, 4].

TABLE 1

Overlap Parameters Describing the Energy Bands of Graphite

Parameter	Description	Range of values (eV)	Value adopted in text (eV)
γ_0	In-plane interaction between nearest neighbors	2 to 4	+3.0
γ_1	Nearest layer interaction between atoms of type a	0.2 to 0.4	+0.3
γ_2	Next nearest layer interaction between atoms of type β	-0.02 to +0.02	-0.018
γ_3	Interlayer interaction between atoms of types a and β producing trigonal warping of bands	0.1 to 0.4	0, 0.3
γ_4	Interlayer interaction producing differences in valence and conduction-band properties	0 to 0.25	+0.2
γ_5	Interlayer interaction producing relative shift of band edges		0
Δ	Shift resulting from difference in a and β atom sites	0 to 0.1	+0.020

B.   The Slonczewski-Weiss Band Model

The results of the calculation by Slonczewski and Weiss [12] can be described using a 4 x 4 sub-Hamiltonian in place of the complete secular determinant (16 x 16).

$$
H = \begin{bmatrix}
\epsilon_1^0 & 0 & H_{13} & H_{13}{}^* \\
0 & \epsilon_2^0 & H_{23} & -H_{23}{}^* \\
H_{13}{}^* & H_{23}{}^* & \epsilon_3^0 & H_{33} \\
H_{13} & -H_{23} & H_{33}{}^* & \epsilon_3^0
\end{bmatrix}, \quad (2)
$$

where

$$\epsilon_1^0 = \Delta + 2\gamma_1 \cos \phi + 2\gamma_5 \cos^2\phi, \tag{3}$$

$$\epsilon_2^0 = \Delta - 2\gamma_1 \cos \phi + 2\gamma_5 \cos^2\phi, \tag{4}$$

$$\epsilon_3^0 = 2\gamma_2 \cos^2 \phi, \tag{5}$$

$$H_{13} = \frac{\sqrt{3}}{2\sqrt{2}} (2\gamma_4 \cos \phi - \gamma_0)a_0\kappa e^{i\alpha}, \tag{6}$$

$$H_{23} = \frac{\sqrt{3}}{2\sqrt{2}} (2\gamma_4 \cos \phi + \gamma_0)a_0\kappa e^{i\alpha}, \tag{7}$$

$$H_{33} = \sqrt{3} \ (\gamma_3 \cos\phi)a_0\kappa e^{i\alpha}, \tag{8}$$

$$\phi = \frac{k_z C_0}{2}, \tag{9}$$

a being the polar angle about the zone edge (Fig. 2).

The energy eigenvalues of the Hamiltonian [Eq. (2)] are then solutions of a fourth-order equation.

$$(\epsilon_1^0 - \epsilon) (\epsilon_2^0 - \epsilon) (\epsilon_3^0 - \epsilon)^2 - [(\epsilon_1^0 - \epsilon) (\epsilon_3^0 - \epsilon) \eta_0^2 (1 + \nu)^2$$

$$+ (\epsilon_2^0 - \epsilon) (\epsilon_3^0 - \epsilon) \eta_0^2 (1 - \nu)^2 + 3 (\epsilon_1^0 - \epsilon) (\epsilon_2^0 - \epsilon) \gamma_3^2 a_0^2 \cos^2 \phi]\kappa^2$$

$$- \sqrt{3}a_0\gamma_3 \cos \phi \cos 3a[(\epsilon_1^0 - \epsilon) \eta_0^2 (1 + \nu)^2 - (\epsilon_2^0 - \epsilon) \eta_0^2 (1 - \nu)^2]\kappa^3$$

$$+ \eta_0^4 (1 + \nu)^2 (1 - \nu)^2 \kappa^4 = 0, \tag{10}$$

where

$$\eta_0 = \frac{\sqrt{3}}{2} \gamma_0 a_0 \quad \text{and} \quad \nu = \frac{2\gamma_4}{\gamma_0} \cos \phi.$$

The variation of the singly degenerate energies ϵ_1^0 and ϵ_2^0 and the doubly degenerate energy ϵ_3^0 along the vertical zone edge HKH' is shown in Fig. 3a.

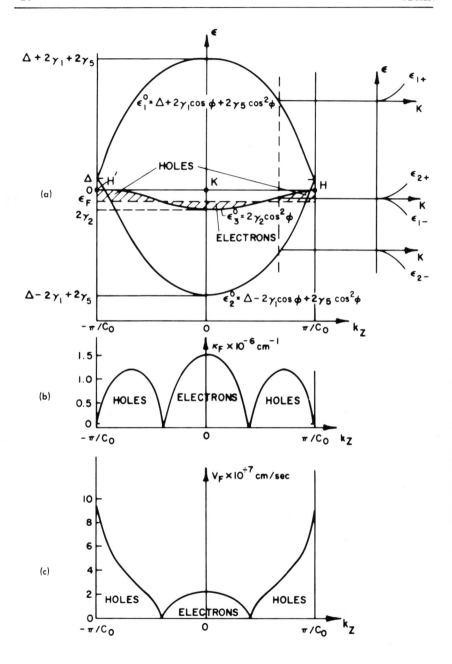

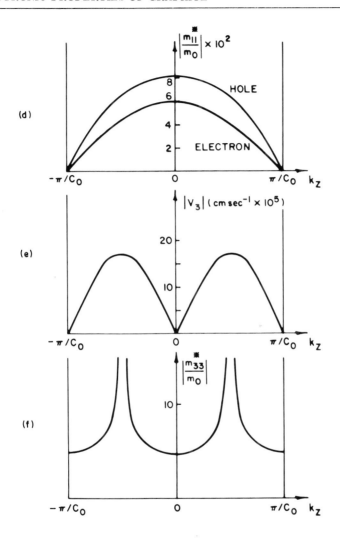

FIG. 3. (a) Variation in the parameters ϵ_1^0, ϵ_2^0, and ϵ_3^0 along the zone edge HKH' ($\varphi = k_z C_0/2$). A negative value of γ_2 and a positive value of Δ are chosen. The variation of energy with the basal component of wave-vector κ is indicated, with states labeled as in the text. At 0°K, electrons from the top of the valence band fill up states at the bottom of the conduction band to

It can be seen that the parameter Δ (resulting from the difference in a and β sites) lifts the degeneracy at the zone corners. The convention used in Fig. 3 is that γ_1 and Δ are positive, but γ_2 is negative, in accord with recent results [18, 20]. For the same value of ϕ the conduction (π^*-antibonding states) and valence bands (π-bonding states) just touch at ϵ_3^0, but the bands overlap for different values of ϕ, with a maximum overlap of $2|\gamma_2|$. Graphite is thus a semimetal with a small number of free electrons occupying the conduction band, with an equal number of holes in the valence band even at 0°K. Without thermal excitation, the energy of the highest filled states in the conduction band coincides with the electrochemical potential, or Fermi level (Fig. 3a).

A negative value of γ_2 implies that electrons occupy states near point K in the Brillouin zone of graphite. Since the dynamical properties of the carriers depend on their position along the zone edge, the assignment of carriers is important in interpreting many phenomena, such as the Hall effect, thermoelectric power, and magnetoreflection. In particular the

---

Fig. 3 - Continued

the Fermi level ϵ_F as shown. (b) Variation in the basal dimension of the Fermi surface as a function of ϕ. The surfaces are not drawn to scale. Since $2\pi/C_0 \simeq 93.7 \times 10^6$ cm^{-1}, the surfaces are highly elongated. (c) The basal component of velocity v_1 for carriers with energy ϵ_F. At the cross-over of the level ϵ_3^0 and ϵ_F ($\phi = \theta$), $\kappa_F = 0$, so that the basal group velocity is zero. (d) The basal component of the force-effective mass m_{11}^* (ϕ). Differences in m_{11}^* for holes and electrons mainly result from the parameter γ_4. Since at low temperatures the holes occupy states for $|\theta| < \phi < \pi/2$, while electrons occupy states for which $-\theta < \phi < \theta$, the average hole effective mass is less than that of the electron. The effective mass is shown for the states along the vertical zone edge ($\kappa = 0$). (e) Variation in carrier-velocity component v_3 parallel to the c axis for electrons along the zone edge. This component is much smaller than the basal component shown in part (c). (f) The c-axis component of the force-effective mass m_{33}^*, which is seen to be approximately $100 m_{11}^*$. Infinite discontinuities in the curves occur at inflection points. Near these points the concept of the force-effective mass is of limited usefulness.

effective-mass component m_{11}^* (see Section III.D) is a maximum at point K, so that a model with negative γ_2 is a "heavy-electron" model.

The assignment of electrons at point K is also favored from simple ideas based on the "nearly free electron" picture of electrons in solids. If one draws the contours of constant electron energy in the zone with increasing energy from the center of the zone, the surfaces would be expected to touch the zone edge at point K before H. It is not surprising therefore that a calculation of the energy bands in graphite [38], using a spherically symmetrical pseudopotential (which gave good agreement for the bands in diamond), predicts that electrons are located at point K. This is in direct contrast to the result of Painter and Ellis [39] who calculated γ_2 directly from a variational calculation using a linear combination of atomic orbitals (LCAO) basis of Bloch states. ($\gamma_2 \sim + 0.02$ eV).

C. The Four-Parameter Band Model

If the parameter $\gamma_3 = 0$, then Eq. (10) can be factorized into two second-order equations

$$\epsilon_{1\pm}(\kappa,\phi) = \frac{\epsilon_1^0 + \epsilon_3^0}{2} \pm \left[\left(\frac{\epsilon_1^0 - \epsilon_3^0}{2} \right)^2 + \frac{3}{4} a_0^2 (\gamma_0 - 2\gamma_4 \cos\phi)^2 \cdot \kappa^2 \right]^{1/2} \tag{11}$$

or

$$\epsilon_{2\pm}(\kappa,\phi) = \frac{\epsilon_2^0 + \epsilon_3^0}{2} \pm \left[\left(\frac{\epsilon_2^0 - \epsilon_3^0}{2} \right)^2 + \frac{3}{4} a_0^2 (\gamma_0 + 2\gamma_4 \cos\phi)^2 \cdot \kappa^2 \right]^{1/2}, \tag{12}$$

giving surfaces of constant energy which are figures of revolution about the axis HKH' (Fig. 4). In the spirit of the $\gamma_3 = 0$ approximation it is reasonable to assume that the parameters γ_4 and γ_5, which are much smaller than γ_0 and γ_1, respectively (Table 1), produce only a small perturbation to the model with $\gamma_4 = \gamma_5 = 0$. The resulting model containing the parameters γ_0, γ_1, γ_2, and Δ is known as the four-parameter band model, with

$$\epsilon_{1,2\pm} = \frac{\epsilon_{1,2}^0 + \epsilon_3^0}{2} \pm \left[\left(\frac{\epsilon_{1,2}^0 - \epsilon_3^0}{2} \right)^2 + \frac{3\gamma_0^2 a_0^2}{4} \kappa^2 \right]^{1/2}, \tag{13}$$

where $\epsilon_{1\pm}$ refers to the expression on the right-hand side with $\epsilon_{1,2}^0 \equiv \epsilon_1^0$; similarly for $\epsilon_{2\pm}$, $\epsilon_{1,2}^0 \equiv \epsilon_2^0$.

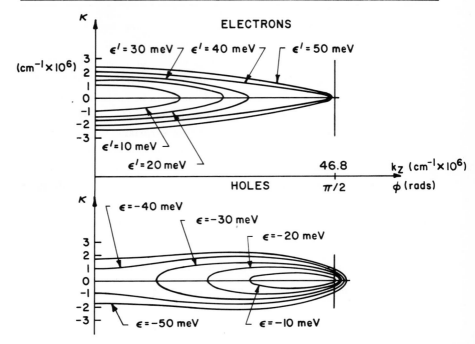

FIG. 4. Constant-energy contours drawn for the electron and hole states, using the four-parameter band model. The surfaces are figures of revolution about the zone edge and are only drawn for the half-zone ($\phi > 0$). Electron-energy surfaces do not touch the hexagonal face of the Brillouin zone for energies (ϵ) less than Δ. The hole surfaces are drawn with the states near the point H extending into the next Brillouin zone for convenience, and the y scale (κ) is magnified five times for convenience. For electrons, ϵ' measures the energy above the bottom of the conduction band ($\epsilon' = \epsilon - 2\gamma_2$).

The four roots of Eq. (13) can be readily identified. Assuming that γ_1 is positive so that ϵ_1^0 is positive and ϵ_2^0 negative away from the zone corners (the solutions near the zone corners are discussed in Section III. F) whereas γ_2 is negative, implying negative values for ϵ_3^0 away from the zone corners, then, as shown in Fig. 3a, the following relationships obtain:

1. $\epsilon_{1+}(\kappa)$ increases as $|\kappa|$ increases and $\epsilon_{1+}(\kappa=0) \equiv \epsilon_1^0$.

2. $\epsilon_{1-}(\kappa)$ decreases as $|\kappa|$ increases and $\epsilon_{1-}(\kappa=0) \equiv \epsilon_3^0$.

3. $\epsilon_{2+}(\kappa)$ increases as $|\kappa|$ increases and $\epsilon_{2+}(\kappa{=}0) \equiv \epsilon_3^0$.

4. $\epsilon_{2-}(\kappa)$ decreases as $|\kappa|$ increases and $\epsilon_{2-}(\kappa{=}0) \equiv \epsilon_2^0$.

This enables the states ϵ_{1+} and ϵ_{2+} to be tentatively labeled as electron-like, and ϵ_{1-} and ϵ_{2-} as holelike (see discussion in Section III.I). It is to be noted that the labeling is dependent on the signs of γ_1 and γ_2. Since $|\gamma_1| \gg |\gamma_2| \simeq |\Delta|$(Table 1), it follows that for temperatures of normal interest for which kT $\leq |\gamma_2|$, the free carriers are restricted to the states labeled ϵ_{1-} and ϵ_{2+} except near the zone corners. The shape of the constant-energy surfaces can be found by inverting the relationship of Eq. (13):

$$\kappa^2 = \frac{4}{3\gamma_0^2 a_0^2} \, [\epsilon^2 - \epsilon(\epsilon_{1,2}^0 + \epsilon_3^0) + \epsilon_{1,2}^0 \cdot \epsilon_3^0]. \tag{14}$$

It is readily seen (Fig. 3b) that the maximum cross section for the electrons occurs at point K ($\phi = 0$) and for holes (from the condition $[\partial(\kappa^2)/\partial\phi]_{\epsilon = \text{const}}$) at $\phi = \psi$, where

$$\cos \psi = \frac{-(\Delta - \epsilon) + \left|(\Delta - \epsilon)^2 + (6\gamma_1^2 \epsilon/\gamma_2)\right|^{1/2}}{6\gamma_1}. \tag{15}$$

$$\therefore \; \cos \psi \simeq \left(\frac{\epsilon}{6\gamma_2}\right)^{1/2}. \tag{16}$$

The values of the extremal cross-sectional area for $\epsilon = \epsilon_F$ are obtained directly from the De Haas-Van Alphen and Shubnikov-De Haas experiments (see Section VI.C) and can be found by substituting for ϕ in Eq. (15). For arbitrary values of ϕ

$$S_{e\ell} \equiv \pi \kappa^2 (\phi) = \frac{4\pi}{3\gamma_0^2 a_0^2} \frac{(2\gamma_2 \cos^2 \phi - \epsilon_F)(\Delta - 2\gamma_1 \cos \phi - \epsilon_F)}{(1 + \nu)^2}, \tag{17}$$

$$S_h = \pi \kappa^2 (\phi) = \frac{4\pi}{3\gamma_0^2 a_0^2} \frac{(2\gamma_2 \cos^2 \phi - \epsilon_F)(2\gamma_1 \cos \phi - \epsilon_F + \Delta)}{(1 - \nu)^2}, \tag{18}$$

where the subscripts "eℓ" and "h" stand for electron and hole, respectively.

The factor $\nu = (2\gamma_4/\gamma_0)\cos\phi$ is included for reasons discussed in Section VI.C. In the four-parameter model ν is strictly zero.

D. The Four-Parameter Model with Parabolic $\epsilon(\kappa)$ Relationship

The dispersion relationships [Eq. (13)] are hyperbolic rather than parabolic, which is the relationship obtained for the nearly-free-electron model of an isotropic conductor. This implies that the force-effective mass (see, for example, Refs. [40] and [41]) in the basal plane ($m_{11}^* = m_{22}^*$) is not constant, but increases as $|\kappa|$ increases, where

$$m_{ij}^* = \hbar^2 \left(\frac{\partial^2 \epsilon}{\partial k_i \partial k_j}\right)^{-1}. \tag{19}$$

For small values of κ the hyperbolic relationship [Eq. (13)] can be expanded:

$$\epsilon_{e\ell} = \epsilon_3^0 + \frac{3\gamma_0^2 a_0^2}{4\left|\epsilon_2^0 - \epsilon_3^0\right|}\kappa^2 + \cdots, \tag{20}$$

$$\epsilon_h = \epsilon_3^0 - \frac{3\gamma_0^2 a_0^2}{4\left|\epsilon_1^0 - \epsilon_3^0\right|}\kappa^2 + \cdots. \tag{21}$$

Then

$$\epsilon_{\pm} \simeq \epsilon_3^0 \pm \frac{3\gamma_0^2 a_0^2}{8\gamma_1 \cos\phi}\kappa^2. \tag{22}$$

The resulting effective mass does not depend on κ in the parabolic band model, but it still varies markedly along the zone edge,

$$m_{\pm} \simeq \frac{4\hbar^2 \gamma_1 \cos\phi}{3\gamma_0^2 a_0^2}. \tag{23}$$

This result is the same as that obtained from the hyperbolic model when $\kappa \to 0$ (Table 2) and is sketched in Fig. 3d for the parameter values given in Table 1.

The formulas for the effective masses of holes and electrons differ only slightly. However, since for low temperatures the electrons are mainly confined to the regions around point K, whereas the holes are located in the "wing" regions, the average basal component of the effective mass of electrons is larger than that for the holes ($<m_{el}>_{av} \simeq 0.06m_0$, $<m_h>_{av} \simeq 0.04m_0$). At temperatures well above the degeneracy temperature ($T_d = \epsilon_F/k$) thermally excited electrons and holes will occupy states along the whole of the edge of the Brillouin zone, and the average masses will be more nearly equal ($\sim 0.05m_0$).

The basal velocity of the carriers can be computed from

$$\underline{v} = \frac{1}{\hbar} \underline{\nabla}_k \epsilon , \tag{24}$$

whence

$$v_1 = \frac{3\gamma_0^2 a_0^2}{2\left|\epsilon_{1,2}^0 - \epsilon_3^0\right|} \kappa = v_2 . \tag{25}$$

This may be compared with the result for the hyperbolic four-parameter model given in Table 2, for which the velocity tends to a constant value for large values of κ, since the dispersion relationship becomes linear for large values of κ. A sketch is given in Fig. 3c of the basal velocity for the parabolic model at the Fermi surface ($\kappa = \kappa_F$) using values for the band parameters given in Table 1.

The parabolic relationship is a reasonably good approximation to the dispersion relationship for carriers near the point K. Near the zone corners the energy varies almost linearly with the wavevector and the approximate formulas of Eqs. (20) and (21) are no longer valid.

It may be seen from Fig. 3d that the effective-mass component m* differs for the conduction and valence bands. This difference arises from the parameter

$$\nu = \frac{2\gamma_4}{\gamma_0} \cos \phi ,$$

which increases the effective mass in the valence band over that in the conduction band for the same value of ϕ, provided that γ_2 is negative.

TABLE 2a

Dispersion Relationship and Principal Velocities
for the Hyperbolic SW Model[a]

Model	Hyperbolic model ($\gamma_3 = 0$)

Dispersion
relationship

$$\epsilon_{1\pm} = \frac{\epsilon_1 + \epsilon_3}{2} \pm \left| \left(\frac{\epsilon_1 - \epsilon_3}{2} \right)^2 + \eta_0^2 (1 - \nu)^2 \kappa^2 \right|^{1/2}$$

$$\epsilon_{2\pm} = \frac{\epsilon_2 + \epsilon_3}{2} \pm \left| \left(\frac{\epsilon_2 - \epsilon_3}{2} \right)^2 + \eta_0^2 (1 + \nu)^2 \kappa^2 \right|^{1/2}$$

Basal velocity

Majority electron

$$\frac{2\eta_0^2}{\hbar a_2} \quad \frac{(1 + \nu)^2}{|\epsilon_2 - \epsilon_3|} \kappa$$

$$v_1 = v_2 = \frac{1}{\hbar} \frac{\partial \epsilon}{\partial \kappa}$$

Majority hole

$$\frac{2\eta_0^2}{\hbar a_1} \quad \frac{(1 - \nu)^2}{|\epsilon_1 - \epsilon_3|} \kappa$$

c-Axis component
of velocity

$$v_3 = \frac{1}{\hbar} \frac{\partial \epsilon}{\partial k_z} = \frac{\gamma_2 C_0}{\hbar} \sin 2\phi$$

[a]Definition of a is given in Table 2e.

TABLE 2b

Dispersion Relationship and Principal Velocity Components for the Parabolic SW, Cylindrical, and Spheroidal Band Models[a]

Parabolic four-parameter model	Cylindrical band model	Spheroidal band model
$\epsilon_{e\ell} = \epsilon_3 + \dfrac{3\gamma_0^2 a_0^2}{4\,|\epsilon_2 - \epsilon_3|}\,\kappa^2$	$\epsilon_{e\ell} = \dfrac{\hbar^2}{2m_{e\ell}}\,\kappa^2$	$\epsilon_{e\ell} = \dfrac{\hbar^2}{2}\left(\dfrac{\kappa^2}{m_1^{e\ell}} + \dfrac{k_3^2}{m_3^{e\ell}}\right)$
$\epsilon_h = \epsilon_3 - \dfrac{3\gamma_0^2 a_0^2}{4\,|\epsilon_1 - \epsilon_3|}\,\kappa^2$	$\epsilon_h = 2\gamma_2 - \dfrac{\hbar^2}{2m_h}\,\kappa^2$	$\epsilon_h = A - \dfrac{\hbar^2}{2}\left[\dfrac{\kappa^2}{m_1^h} + \dfrac{(k_3 - k_0)^2}{m_3^h}\right]$
$\dfrac{2\eta_0^2}{\hbar\,|\epsilon_2 - \epsilon_3|}\,\kappa$	$\dfrac{\hbar}{m_{e\ell}}\,\kappa$ for electrons	$\dfrac{\hbar}{m_1^{e\ell}}\,\kappa$ for electrons
$\dfrac{2\eta_0^2}{\hbar\,|\epsilon_1 - \epsilon_3|}\,\kappa$	$\dfrac{\hbar}{m_h}\,\kappa$ for holes	$\dfrac{\hbar}{m_1^h}\,\kappa$ for holes
$\dfrac{\gamma_2 C_0}{\hbar}\,\sin 2\phi$	Zero	$\dfrac{\hbar}{m_3^{e\ell}}\,k_3$ for electrons
		$\dfrac{\hbar}{m_3^h}\,(k_3 - k_0)$ for holes

[a]For definitions of A and k_0, see Eq. (35) and Table 2d, respectively.

TABLE 2c

Force Effective Mass and Extremal Area Equations for the Hyperbolic SW
Band Model[a]

Model	Hyperbolic model ($\gamma_3 = 0$)
Basal component of force effective mass $m^*_{1\,1} = m^*_{2\,2}$ $= \dfrac{\hbar^2}{\partial^2 \epsilon / \partial \kappa^2}$	Majority electron $\dfrac{\hbar^2 \left\| \epsilon_2 - \epsilon_3 \right\|^{a_2}}{2\eta_0^2 (1 + \nu)^2} \quad \dfrac{1}{1 - \beta_2}$ Majority hole $\dfrac{\hbar^2 \left\| \epsilon_1 - \epsilon_3 \right\|^{a_1}}{2\eta_0^2 (1 - \nu)^2} \quad \dfrac{1}{1 - \beta_1}$
c-Axis component of force effective mass $m^*_{33} = \dfrac{\hbar^2}{\partial^2 \epsilon / \partial k_z^2}$	$\dfrac{\hbar^2}{\gamma_2 C_0^2 \cos 2\phi}$
Extremal area at point K ($S_{e\ell}$)	$\dfrac{\pi}{\eta_0^2 (1 + \nu_0)^2} \left[\epsilon_F^2 - \epsilon_F (\Delta - 2\gamma_1 + 2\gamma_5 + 2\gamma_2) \right.$ $\left. + 2\gamma_2 (\Delta - 2\gamma_1 + 2\gamma_5) \right]$
Position of hole extremal area	$\cos \psi \simeq \left(\dfrac{\epsilon_F}{6\gamma_2} \right)^{1/2}$

[a]Definition of β is given in Table 2e.

TABLE 2d

Force Effective Mass and Extremal Area Equations for Parabolic SW,
Cylindrical, and Spheroidal Band Models

Parabolic four-parameter model	Cylindrical band model	Spheroidal band model
$\dfrac{\hbar^2}{2\eta_0^2}\,\lvert \epsilon_2 - \epsilon_3\rvert$	$m_{e\ell}$ for electrons	$m_1^{e\ell}$ for electrons
$\dfrac{\hbar^2}{2\eta_0^2}\,\lvert \epsilon_1 - \epsilon_3\rvert$	m_h for holes	m_1^h for holes
$\dfrac{\hbar^2}{\gamma_2 C_0^2\,\cos 2\phi}$	Infinite	$m_3^{e\ell}$ for electrons m_3^h for holes
$\dfrac{\pi}{\eta_0^2}\,(\epsilon_F - 2\gamma_2)\cdot$ $(2\gamma_1 - \Delta + 2\gamma_2)$ $= \dfrac{2\pi\lvert\gamma_2\rvert}{3\eta_0^2}\cdot$ $(2\gamma_1 - \Delta + 2\gamma_2)$	$\dfrac{2\pi m_{e\ell}\epsilon_F}{\hbar^2}$	$\dfrac{2\pi m_1^{e\ell}\epsilon_F}{\hbar^2}$
$\cos\psi = \left(\dfrac{\epsilon_F}{6\gamma_2}\right)^{1/2}$ $= \left(\dfrac{2}{9}\right)^{1/2}$	Independent of k_3	$k_z = k_0 = \dfrac{2\psi}{C_0}$

TABLE 2e

Extremal Area for Majority and Minority Holes and Definitions for α and β Which Appear in Tables 2a and 2c, for the Hyperbolic SW Model.

Model	Hyperbolic model ($\gamma_3 = 0$)	
Hole extremal area (S_h)	$\dfrac{\pi}{\eta_0^2 (1 - \nu)^2} \Big\lvert \epsilon_F^2 - \epsilon_F (\Delta + 2\gamma_1 \cos \psi + 2\gamma_5 \cos^2 \psi$ $+ 2\gamma_2 \cos^2 \psi)$ $+ 2\gamma_2 \cos^2 \psi (\Delta + 2\gamma_1 \cos \psi + 2\gamma_5 \cos^2 \psi \Big\rvert$	
Extremal cross section for planes $\phi = \pm \pi/2$	$\dfrac{\pi \epsilon_F (\epsilon_F - \Delta)}{\eta_0^2}$	
Relationship between ϵ_F and band overlap	Not simple	Approximately equal to $1.2 \gamma_2$ for currently accepted values of the band parameters

$$\phi = \frac{k_z C_0}{2}; \quad \eta_0^2 = \frac{3\gamma_0^2 a_0^2}{4}; \quad \nu = \frac{2\gamma_4}{\gamma_0} \cos \phi; \quad a_1 \equiv \left\lvert \left\lvert 1 + \left(\frac{2}{\epsilon_1 - \epsilon_3}\right)^2 \eta_0^2 (1-\nu)^2 \kappa^2 \right\rvert \right\rvert^{\frac{1}{2}};$$

$$\beta_1 = \frac{4\eta_0^2 (1-\nu)^2 \kappa^2}{\lvert \epsilon_1 - \epsilon_3 \rvert^2 a_1^{3/2}}; \quad \beta_2 = \frac{4\eta_0^2 (1+\nu)^2 \kappa^2}{\lvert \epsilon_2^0 - \epsilon_3^0 \rvert^2 a_2^{3/2}}; \quad a_2 \equiv \left\lvert \left\lvert 1 + \left(\frac{2}{\epsilon_2 - \epsilon_3}\right)^2 \eta_0^2 (1+\nu)^2 \kappa^2 \right\rvert \right\rvert^{\frac{1}{2}};$$

$$\epsilon_1 \equiv \epsilon_1^0; \quad \epsilon_2 \equiv \epsilon_2^0; \quad \epsilon_3 \equiv \epsilon_3^0$$

TABLE 2f

Extremal Areas for Majority and Minority Holes for the Parabolic SW, Cylindrical, and Spheroidal Band Models. The Relationship for ϵ_F is also given

Four-parameter model	Cylindrical band model	Spheroidal band model
$\dfrac{8\pi}{9\eta_0^2} \,\lvert \gamma_2 \rvert \cdot$ $\left[\Delta + \left(\dfrac{8}{9}\right)^{1/2} \gamma_1 - \dfrac{4}{9}\gamma_2\right]$	$\dfrac{2\pi m_h \,(2\gamma_2 - \epsilon_F)}{\hbar^2} = S_{e\ell}$	$\dfrac{2\pi m_1^h}{\hbar^2}\,(A - \epsilon_F)$
$\dfrac{-\pi_F \Delta}{\eta_0^2}$	Not applicable	Separate spheroids may be specified for the minority carriers, centered on point H
$\epsilon_F = \dfrac{4\,\gamma_2}{3}$	$\epsilon_F = 2\gamma_2 \dfrac{m_h}{(m_{e\ell} + m_h)}$	$\dfrac{\epsilon_F^3}{(2\gamma_2 - \epsilon_F)^3} = \dfrac{(m_1^2 m_3)^h}{(m_1^2 m_3)^{e\ell}}$

E. Trigonal Warping of the Bands Due to γ_3

The effect of a nonzero value for the parameter γ_3 is mainly to introduce a modulation of the constant-energy surfaces with the polar angle a. It is not surprising that the crystal symmetry imposes three-fold rotation symmetry, or trigonal warping (Fig. 5). Though the full details of the bands can be obtained by solving directly the secular Eq. (2), the warping may be included as a perturbation [43] provided that $\epsilon_{1,2}^0$ is well separated from ϵ_3^0. Using a parabolic expression for the energy-wavevector relationship before γ_3 is included, we can compare directly with Eq. (22):

$$\epsilon - \epsilon_3^0 \simeq \frac{3a_0^2\gamma_0^2}{4\left|\epsilon_2^0 - \epsilon_3^0\right|}\ \kappa^2 \pm \frac{\sqrt{3}}{4}\,a_0\gamma_3 \cdot \kappa\cos\phi\left(1 + \frac{2\kappa}{\kappa_s}\sin 3a + \frac{\kappa^2}{\kappa_s^2}\right)^{1/2}, \quad (26)$$

where

$$\kappa_s = \frac{2\sqrt{3}\gamma_1\gamma_3}{a_0\gamma_0^2}\ \cos^2\phi. \qquad (27)$$

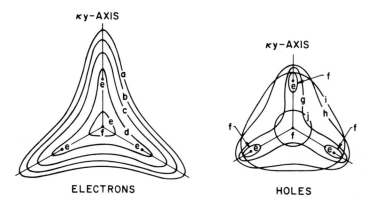

FIG. 5. Cross sections of the Fermi surface of graphite indicating the trigonal warping due to γ_3, taken to be approximately 0.2 eV in the diagram. In units of $\pi/2$, ϕ is (a) 0, (b) 0.18, (c) 0.27, (d) 0.36, (e) 0.408, (f) 0.447, (g) 0.5, (h) 0.625, (i) 0.75, (j) 1.0. After Ono and Sugihara [42].

The term κ_s gives an estimate of the range of κ over which trigonal warping is effective. It can also be seen that the warping is more pronounced for $\cos \phi \simeq 1$, that is, near point K, until at point H the surfaces have a circular cross section. A sketch of the Fermi surface is given in Fig. 6. It can be seen that at the joining of the hole and electron surfaces there are three satellite, or "outrigger," pieces. There is no direct experimental evidence of these pieces at the moment, although the presence of trigonal warping has been reasonably established by cyclotron-resonance and magnetroflection experiments. The latter measurements suggest that γ_3 is of such magnitude (0.3 eV) [45] that perturbation treatments do not adequately describe the energy bands and that warping is more severe than is indicated in Figs. 5 and 6 (see Section VI. C).

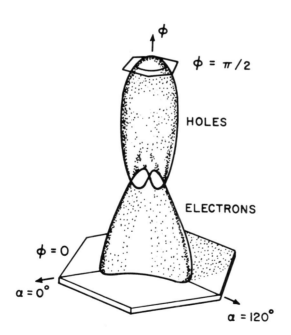

FIG. 6. The Fermi surface of graphite showing the trigonal warping introduced by the parameter γ_3. The surface is only shown for positive values of ϕ. The conical tip of the hole surface is placed outside the first Brillouin zone for convenience. After Dresselhaus and Mavroides [44].

$G^{\neq 1, 2\gamma_2}$

F. Details of the Bands near the Zone Corners

If the parameter $\Delta > \epsilon_F$ for $\gamma_2 < 0$ (Table 1), then the Fermi surface touches the hexagonal face of the Brillouin zone, as indicated by experiments (see Section VI.B). Since Δ is very small, the singly degenerate level ϵ_2^0 crosses the doubly degenerate level ϵ_3^0 very close to the zone edge. With notation used in Fig. 7,

$$\phi_c = \cos^{-1}\left(\frac{\Delta}{2\gamma_1}\right) \tag{28}$$

since $\epsilon_3^0 = 0$, so that

$$\delta k_c = \left[\frac{\pi}{C_0} - (k_z)_{cr}\right] \simeq \frac{\Delta}{\gamma_1 C_0}, \tag{29}$$

where the subscript "cr" stands for crossover.

It is instructive to study the solutions for the four-parameter band model more closely near the zone corners. If Δ is positive, then the states labeled ϵ_{1-} [Eq. (11)] comprising the majority holes, are not associated with a discontinuity at ϕ_c and have a nonzero value of κ_F that varies smoothly to the zone corner. The resulting Fermi surface then touches the hexagonal face of the Brillouin zone, with the cross section at this plane obtained from Eq. (14).

$$S = \frac{4\pi}{3\gamma_0^2 a_0^2} \epsilon_F(\epsilon_F - \Delta), \tag{30}$$

which depends sensitively on the parameter Δ.

The states labeled ϵ_{1+} similarly are not associated with a discontinuity, whereas the states associated with the level ϵ_2^0, namely, ϵ_{2+} and ϵ_{2-}, need to be considered carefully since ϵ_2^0 crosses ϵ_3^0. For these states the labeling changes at $\phi = \phi_c$ as indicated in Fig. 7. Thus the energy <u>decreases</u> as $|\kappa|$ increases for <u>both</u> bands for which $\epsilon(\kappa = 0) = \epsilon_3^0$, for $|\phi_c| < |\phi| < \pi/2$ and increases in this region for the band for which $\epsilon(\kappa = 0) = \epsilon_2^0$ (cf. Section III.c).

FIG. 7. Variation in principal energies near the zone corners: (a) Δ positive; (b) $2\gamma_2 < \Delta < 0$; (c) $\Delta < 2\gamma_2$. The states are labeled as in the text, and the variation in energy with the basal component of the wavevector (κ) indicated. Here ϕ_c represents the crossover of ϵ_3^0 and ϵ_1^0 or ϵ_2^0, and ϕ_c' the crossover of ϵ_F and ϵ_1^0 or ϵ_2^0.

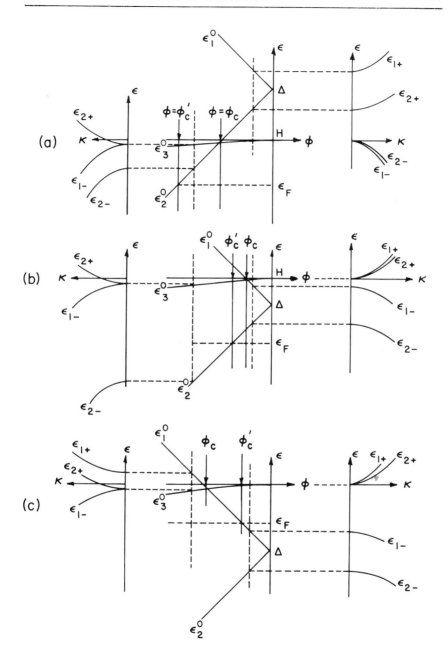

This gives an extra pocket of holelike states in the corner of the zone for which

$$\epsilon_{max} = \epsilon(\kappa = 0) = \epsilon_3^0 \quad \text{for} \quad |\phi_c| < |\phi| < \pi/2,$$

$$\epsilon_{max} = \epsilon(\kappa = 0) = \epsilon_2^0 \quad \text{for} \quad |\phi'_c| < |\phi| < |\phi_c|,$$

where ϕ_c' is the crossover point of ϵ_2^0 and ϵ_F (Fig. 7a).

The properties of the carriers in this region are only slightly modified by the parameter γ_3, and so are figures of revolution. The dispersion relationship is unusual. At $\phi = \phi_c$, the relationship $\epsilon(\kappa)$ is linear,

$$\epsilon = \frac{\sqrt{3}}{2}\gamma_0 a_0 \kappa, \tag{31}$$

which agrees with the result of Wallace [4] for two-dimensional graphite near the zone corners, when allowance is made for the difference in the definition of the wavevector κ employed. For points away from $\phi = \phi_c$ the dispersion relationship is again hyperbolic, but since $\left|\epsilon_{1,2}^0 - \epsilon_3^0\right|$ is small, the $\epsilon(\kappa)$ relationship becomes asymptotic to a linear form for relatively small values of κ ($\sim \kappa_F$).

Figure 8 sketches the shape of the Fermi surface near point H in the extended-zone scheme. In sketches of the complete Fermi surface for graphite it is conventional to keep the majority electron and hole surfaces in the first Brillouin zone and to sketch the minority surface outside this zone as a "nose cone" to the "rocketlike" majority surfaces (Fig. 6).

G. Effects of Spin-Orbit Interaction

A coupling exists between the magnetic field produced by the orbital motion of electrons and the electronic magnetic moment. The Hamiltonian including the effect of this spin-orbit interaction has been calculated [46]. In the carbon atom the magnitude of this interaction would be expected to be on the order of 8×10^{-3} eV, which is comparable with the band overlap. However, McClure and Yafet [47] showed that the effect of the graphite lattice reduces the interaction to about 2×10^{-4} eV, so that the Slonczewski-Weiss band model without spin-orbit interaction is a very good first approximation.

The main effect of the interaction is to lift degeneracies in the bands. Dresselhaus and Dresselhaus [46] were able to show that if the k_3 dependence of the effect were neglected, it could be discussed by using only three par-

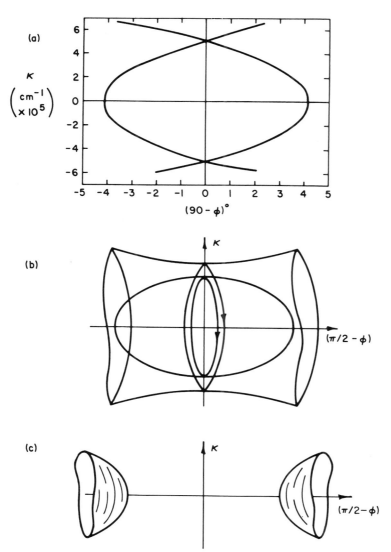

FIG. 8. The Fermi surface near point H, with negative γ_2 in all cases.
(a) $\Delta > 2\gamma_2$. Spin-orbit interaction is not included. The surfaces are shown
extending into the next Brillouin zone (extended-zone scheme). Actual values
for κ_F are given for $\Delta = 20$ and $\epsilon_F = -22$ meV. (b) $\Delta > 2\gamma_2$. Spin-orbit
interaction produces a pocket of minority holes. Both the majority and minority-carrier surfaces are extremal in the plane $\phi = \pi/2$. (c) $\Delta < 2\gamma_2$.
The Fermi surface does not extend as far as the hexagonal face of the
Brillouin zone.

ameters $(\lambda_{1,2}, \lambda_{1,3}, \lambda_{3,3})$. The parameters $\lambda_{1,2}$ and $\lambda_{3,3}$ are approximately equal (~2 x 10^{-4} eV), but they are much larger than $\lambda_{1,3}$, which will be neglected in the following discussion. At point K the doubly degenerate level ϵ_3^0 is split into two levels with energy $\lambda_{3,3}$ above and below the unperturbed level, respectively (Fig. 9a). The nondegenerate levels ϵ_1^0 and ϵ_2^0 are raised and lowered, respectively, by an amount $(\lambda_{1,2})^2/4\gamma_1$. At point H the degeneracy of the levels ϵ_1^0 and ϵ_2^0 is lifted, with energy separation $2\lambda_{1,2}$ (Fig. 9b). The lifting of this degeneracy has a particularly important effect on the shape of the Fermi surface near point H when spin-orbit interaction is included. The minority-carrier surface is separated from the majority surface as shown in Fig. 8b. The resulting change in the cross section $\delta S/S_0$ amounts to only about 1%, if $\lambda_{3,3} \sim 10^{-4}$ eV, which is too small to be observed experimentally (see Section VI.D), but most importantly (a) the cross-sectional areas and (b) the effective masses are stationary with respect to k_z at $k_z = \pm \pi/C_0$, so that the carriers there may be observed in (a) De Haas-Van Alphen and (b) cyclotron-resonance experiments, respectively.

The exact magnitudes of the spin-orbit splitting parameters are not known exactly at present, but it is well established that they are less than 10^{-3} eV (see also Section VI.F).

H. Simple Models for the Energy Bands in Graphite

Since the interaction between the layers is so much weaker than interactions between atoms in the plane, graphite can be considered to a first approximation as a two-dimensional solid. Johnston [49] proposed that the bands could be approximated by cylinders extending from $\phi = -\pi/2$ to $+\pi/2$ with a parabolic energy-wavevector relationship:

$$\epsilon_{el} = \frac{\hbar^2}{2m_{el}} \kappa^2, \tag{32}$$

$$\epsilon_h = \epsilon_0 - \frac{\hbar^2}{2m_h} \kappa^2, \tag{33}$$

where $\epsilon_0 \sim |2\gamma_2|$. Such a band model allows basal transport phenomena to be calculated very simply; it has been used extensively by Klein [50].

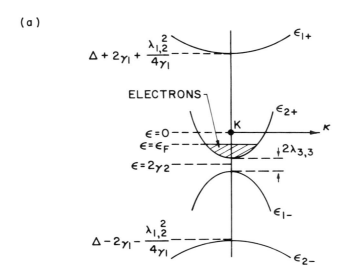

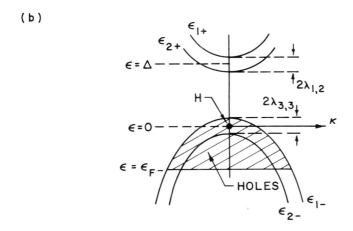

FIG. 9. The effects of spin-orbit interaction on the bands for states in
(a) the plane $\phi = 0$ and (b) the planes $\phi = \pm \pi/2$. In (a) states ϵ_{1+} and ϵ_{2-}
are moved apart as shown, while the splitting of the doubly degenerate levels
ϵ_{2+} and ϵ_{1-} is exaggerated for clarity. In (b) spin-orbit splitting of the doubly
degenerate ϵ_{1-} and ϵ_{2-} levels results in the separation of hole states into a
majority-hole surface and minority-hole pocket. The degenerate states
ϵ_{1+} and ϵ_{2+} are also split.

This model assumes that the carrier velocity in the c-axis direction is zero and replaces the spectrum of properties obtained in the Slonczewski-Weiss model with discrete properties. As can be seen from Fig. 4, the constant-energy surfaces in the Slonczewski-Weiss model somewhat resemble cylinders for values of κ larger than κ_F.

The properties of carriers for ellipsoidal bands are well established (see, for example, Ref. [51]). Since the constant-energy surfaces in graphite resemble elongated cigars with axial ratios of approximately 15:1, it is sometimes convenient to represent the dispersion relationships with a spheroidal model ($m_{11} = m_{22} \neq m_{33}$):

$$\epsilon_{e\ell} = \frac{\hbar^2}{2} \left[\frac{\kappa^2}{m_1^{e\ell}} + \frac{k_z^2}{m_3^{e\ell}} \right], \tag{34}$$

$$\epsilon_h = A - \frac{\hbar^2}{2} \left[\frac{\kappa^2}{m_1^h} + \frac{k_z^2}{m_3^h} \right], \tag{35}$$

where A is the band overlap, $A \sim 2\gamma_2 (1-\cos^2 \psi)$. Such surfaces can be fitted along the vertical edge of the Brillouin zone (Fig. 10a), and the minority carrier can also be included with a separate spheroid. The model is convenient for calculating properties at low temperatures. At higher temperatures the standard formulas include carriers that are unrestricted in k space. This allows holes outside the first Brillouin zone (Fig. 10b) to contribute to electronic properties, thereby introducing noticeable errors in the variation of the number of hole carriers and shift of the electrochemical potential with temperature.

This model also allows the density of states for the carriers to vary with the square root of the energy relative to the band edge. As indicated in Section V.A. this relation holds approximately for the conduction-band states near the electrochemical potential but does not adequately describe the variation for valence-band states. The model therefore is of limited usefulness, particularly in predicting thermoelectric and thermomagnetic properties (see Sections VII.I and VII.J).

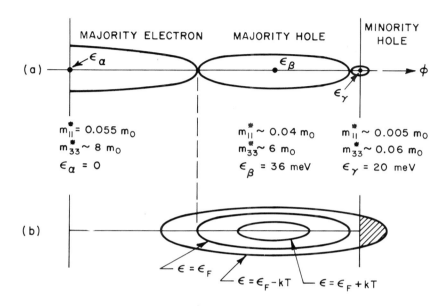

FIG. 10. (a) The Fermi surface for the spheroidal model of graphite, including minority carriers. This model can be characterized by eight independent parameters: two principal effective-mass components for each carrier type, and two band-overlap parameters for majority and minority carriers. Typical values are shown. The electrochemical potential (Fermi energy) can then be calculated from the condition of charge neutrality. (b) Standard formulas for spheroidal bands do not account for restrictions on allowed k values. Hence electron and hole states outside the first Brillouin zone (shaded region) are counted. At temperatures above the degeneracy temperature ($T_d = \epsilon_F/k$) this leads to errors in the calculated number of carriers. Surfaces are only drawn for the hole states for clarity.

I. Labeling of States in Graphite

It is convenient at this point to illuminate the convention used to label the states as electronlike or holelike. If the energy is a quadratic function of the wavevector, then it is convenient to think of bands of normal or inverted form, corresponding to electron bands and hole bands, respectively (see, for example Ref. [52]). This assumes, however, that the derivatives $\partial^2 \epsilon / \partial k^2$ are of the same sign in all directions. It is possible to envisage a case in which one component of the derivative $\partial^2 \epsilon / \partial k_z^2$ is negative, with $\partial^2 \epsilon / \partial k_x^2$ and $\partial^2 \epsilon / \partial k_y^2$ positive. In this case the labeling of states as holelike or electronlike becomes somewhat confusing.

The dynamics of an electron in a uniform, constant, applied field $\underline{F}$ is given by the equation:

$$\hbar \frac{dk}{dt} = -e\underline{F}, \tag{36}$$

so that the wavevector $\underline{k}$ changes at a uniform rate between collisions. For the field in directions x and y in our example the wavevector changes so that the electron velocity increases in the direction of the applied force $-e\underline{F}$, whereas in the third direction the velocity will decrease.

Clearly, a map of the variation in electron energy with wavevector as a function of k_x and k_z or k_y and k_z would eventually give rise to a saddle point. Figure 11 illustrates such a saddle point H. At 0°K the electrons mostly occupy regions of the bands where the curvature is sensibly of the same sign for all directions. At higher temperatures ($kT \geq \epsilon_F$) electrons will occupy states near point H, and holes will occupy states near point K, where the labeling is not unambiguous.

Similar difficulties are encountered in metals and semiconductors for states near the Brillouin-zone boundaries (see, for example, Ref. [40]), which delineate surfaces for which the diffraction condition

$$2\underline{k} \cdot \underline{G} + G^2 = 0 \tag{37}$$

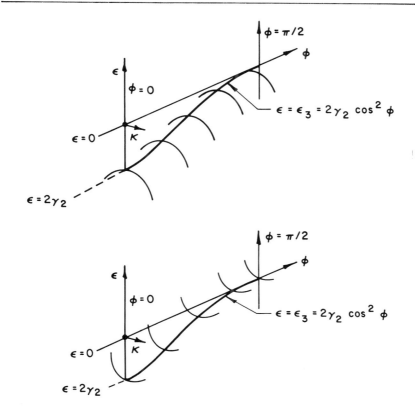

FIG. 11. Variation in energy with wavevector components κ and k_z = $2\phi/C_0$. Top: states for which $\partial\epsilon/\partial\kappa$ is negative (holelike); Near $\phi = \pi/2$, $\partial\epsilon/\partial k_z$ is also negative, but near $\phi = 0$, $\partial\epsilon/\partial k_z$ is positive, so that the labeling of states is ambiguous. Bottom: states for which $\partial\epsilon/\partial\kappa$ is positive. In this case the labeling is ambiguous near $\phi = \pi/2$.

is satisfied, $\underline{G}$ being a reciprocal lattice vector. The waves for $\underline{k} = \underline{G}$ are standing waves with $\underline{v} = 0$. It is not surprising therefore that the acceleration of electrons near the zone boundaries is strongly influenced by the interaction of the electron waves with the crystal.

It is conventional in graphite to label the states with positive $\partial\epsilon/\partial\kappa$ as electron-like and those with negative $\partial\epsilon/\partial\kappa$ as hole-like and this procedure will be adopted here.

J. Effect of Correlation of Electron Motion Due to Coulomb Interaction

The Slonczewski-Weiss band model is based on the one-electron approximation, in which it is assumed that each electron moves in a static, periodic potential due to the ion cores and the smeared-out effects of all the other electrons. McClure [17] has pointed out that the correlations between electrons due to their Coulomb interaction should be important in graphite. The theoretical treatment of such effects is difficult and has only in the last few years received adequate attention (see Refs. [53] and [54]).

Many-body effects are usually discussed in terms of collective vibrations of the gas of electrons about the lattice of positive ions. The natural frequency of the compressional vibration, known as the plasma frequency, is given by

$$\omega_p = \left(\frac{4\pi n e^2}{m^* K}\right)^{1/2}, \tag{38}$$

where n is the free-electron density, m* is the effective mass, and K is an effective dielectric constant. **Energy eigenvalues of this vibrational mode have allowed values,**

$$\epsilon_p = (\ell + 1/2)\hbar\omega_p, \tag{39}$$

where ℓ is an integer and $1/2\hbar\omega_p$ is the ground-state energy [53, 55]. Although it strictly applies to an isotropic electron gas with constant value for m*, Eq. (38) can be used to estimate the plasma energy in graphite. For motion parallel to the layer planes, $\hbar\omega_p^{(1)} \simeq 0.17$ eV at 0°K, whereas for motion perpendicular to the planes, $\hbar\omega_p^{(2)} \simeq 0.01$ eV [17], so that $\hbar\omega_p^{(1)}/\epsilon_F \simeq 10$, $\hbar\omega_p^{(2)}/\epsilon_F \simeq 1$. It can be shown [53] that for an isotropic electron gas for which $\hbar\omega_p/\epsilon_F$ is greater than 4, the Coulomb interaction is so great that the ground state corresponds to an electron solid. McClure [17] therefore tentatively suggests that the electrons in graphite behave like an electron liquid, in which case the Slonczewski-Weiss band model describes the quasiparticle energy spectrum [53]. McClure further suggests that this model should account adequately for the De Haas-Van Alphen effect and electronic heat capacity, but it may give rise to discrepancies in the interpretation of cyclotron resonance and the steady diamagnetic susceptibility.

Linderberg and Makila [9] and Linderberg [8] have attempted to assess the importance of correlation effects for the π electrons in single and weakly interacting graphite layers, using a model for the Hamiltonian proposed by Hubbard [54]. Their result showed that the density of states was not appreciably altered by the introduction of the correlation correction, but that dynamical properties (damping) of the electrons with energies far from the Fermi energy were appreciably modified.

An attempt to describe the effects of anisotropy on the electronic correlations in graphite has been made in a model anisotropic system in which electrons were constrained to move on parallel, equally spaced planes [56]. The screening properties of the system were then investigated by calculating the dielectric response of the system to an external point charge on one of the planes. In an isotropic metal the potential of a charge q is screened by the free carriers, of density n (cm^{-3}). The screened potential can be computed by using the Thomas-Fermi approximation as [53, 55, 57]

$$V(r) = \frac{q}{r} e^{-r/\lambda},\tag{40}$$

where

$$\lambda = \left(\frac{\epsilon_F}{6\pi ne^2}\right)^{1/2} = \text{screening length.}\tag{41}$$

The screening length is therefore a measure of the range of the potential. Visscher and Falicov [56] show that the screening is relatively unimportant if only the free-electron density $(10^{-4}$ electron per atom) is used in the calculation. With the entire π-electron density included, the screening distance becomes very small $(\sim 0.5 \overset{\circ}{A})$, but it is not appreciably different along and perpendicular to the planes. As shown in Fig. 12, the Thomas-Fermi approximation [Eq. (40)] does not describe the potential adequately, since, in the self-consistent treatment (random-phase approximation) Friedel oscillations appear [58]. These oscillations come about from the redistribution of charge around the test charge q. In the case discussed by Visscher and Falicov [56] a diffuse ring of induced charge is created in the same plane as the test charge.

A similar calculation has been reported by Boardman and Graham [59], utilizing the SW band model for a description of the effective masses, and is discussed in greater detail in Section VII.C. In this treatment the

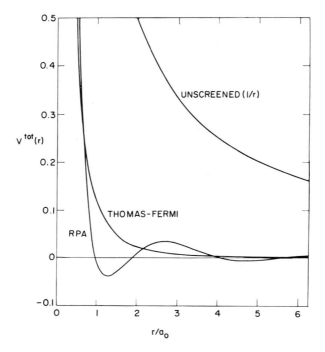

FIG. 12. Comparison of the Coulomb potential q/r with the screened potential $(q/r)e^{-r/\lambda}$ (Thomas-Fermi approximation) and the self-consistent potential obtained by using the random-phase approximation (RPA). The abscissa is expressed in terms of multiples of the Bohr radius (0.53 Å). Curves are drawn for the problem of parallel sheets of charge discussed in the text for the complete π-electron density. Friedel oscillations can be seen for the curve RPA. Reprinted from Ref. [56] by courtesy of the American Institute of Physics.

self-consistent potential from point test charges (ionized impurities) was described in terms of a dielectric response function $\epsilon(\underset{\sim}{q})$ [60]:

$$V(\underset{\sim}{q}) = \frac{V_0(\underset{\sim}{q})}{\epsilon(\underset{\sim}{q})} , \qquad (42)$$

where $V(\underset{\sim}{q})$ is the Fourier transform of the screened potential and $V_0(\underset{\sim}{q})$ is that of the unscreened potential. The dielectric function was shown to be the sum of an isotropic $\epsilon_i(\underline{q})$ and an anisotropic part $\epsilon_a(q_{||}, q_\perp)$, where $q_{||}$ and $q_\perp$ are the components of the reciprocal lattice vector parallel and perpendicular to the planes, respectively. It was further shown that the anisotropic part of $\epsilon(\underline{q})$ is very important in modifying the electron-impurity collisions (see Section VII. C). The results of Boardman and Graham [59] would therefore appear to be in conflict with those of Visscher and Falicov [56].

At present the modification of electron properties of carriers near the Fermi energy introduced by correlation effects is not fully understood. It is possible that such effects are important in controlling the c-axis conductivity. This is discussed in Section VIII.

K. Other Calculations of the Energy Bands in Graphite

Van Haeringen and Junginger [39] have calculated the energy bands for high symmetry directions using a pseudopotential approach. (For an explanation of the pseudopotential method see Ref. [61].) As already stated in Section III. B, their method predicted that electrons should occupy states near point K. The method may be criticized on the grounds that a local, centrally symmetrical pseudopotential was used, that had previously given reasonable results for the energy-band structure in diamond and silicon carbide. This is probably not realistic for the anisotropic situation of graphite. The results of this theory were not expressed in terms of the overlap parameters used in the theory of Slonczewski and Weiss [12]. However, the values for the parameters γ_1 (0.4 eV), γ_2 (-0.0425 eV), and Δ (0.415 eV) may be compared directly with the values obtained by other methods (see Section VI. I). In particular, the values for γ_2 and Δ are much larger than other estimates.

Other calculations of the band energies along high-symmetry directions in wavevector space have been reported for the two-dimensional [62] and three-dimensional [39, 63] structure. These calculations were mainly concerned with fitting optical data, and hence will not be considered in detail here.

Haering [30] reported a calculation of the energy bands near the vertical edges of the Brillouin zone for rhombohedral graphite, obtained by using the nearest-neighbor tight-binding approximation. His results showed that the two π bands touched for points no longer lying along the zone edges, but lying on cylinders with axes coincident with the edges. McClure [31] extended the calculation for next-nearest neighbors and found that the Fermi surface would resemble a string of very long "link sausages" wound in a loose spiral about the zone edge. For this structure the density of states was calculated to be only about 7% of that for the Bernal structure, giving reasonable agreement with susceptibility measurements on rhombohedral graphite obtained by crushing.

IV. BAND THEORY OF GRAPHITE IN NONZERO MAGNETIC FIELDS

A. Introduction

In the preceding section we computed the properties of the electrons for zero magnetic field, for which the states were labeled with the wavevector $\underline{k}$. In nonzero magnetic fields it is necessary to label the states with a magnetic quantum number (m). Classically, when a field H is applied to a free-electron gas, circular motion is described in a plane perpendicular to the field direction, with the angular frequency, called the cyclotron frequency ω_c, given by

$$\omega_c = \frac{eH}{m_0 c} .$$
(43)

When the effect of the crystal potential is taken into account, this expression is modified by replacing the free-electron mass m_0 with the cyclotron-effective mass m_c^*. m_c^* is approximately equal to the force-effective m^*.

The permitted orbits are quantum mechanically those that enclose half-multiples of a quantum of flux $2\pi\hbar c/e$ and have orbital areas A in real space, given [64] by

$$A_m = (m + \frac{1}{2}) \frac{2\pi\hbar c}{eH}$$
(44)

and orbital areas in reciprocal space, S, given by

$$S_m = (m + \frac{1}{2}) \frac{2\pi eH}{\hbar c} .$$
(45)

This result holds for general orbital shapes. As the orbit is quantized, the energy also may assume values:

$$\epsilon_m = (m + \frac{1}{2}) \hbar\omega_c$$
(46)

for a two-dimensional electron gas (no motion parallel to the field direction). For the three-dimensional case of electrons in solids, the energy will vary with particle velocity in the field direction. We can imagine the allowed states in reciprocal space to lie on tubes of constant magnetic energy (Fig. 13b), each tube having a degeneracy depending on the magnetic field. The mean density of states in reciprocal space is not expected to vary too much from

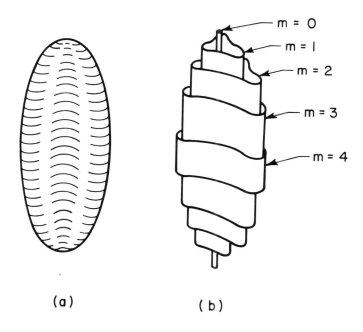

(a)   (b)

FIG. 13. The effect of a magnetic field on the allowed states in reciprocal space. (a) H = 0. Allowed values of the wavevector $\underline{k}$ are distributed evenly throughout reciprocal space, with each allowed state of one spin direction occupying a volume $8\pi^3/V$, where V is the volume of the crystal. (b) H ≠ 0. The wavevector $\underline{k}$ is no longer a good quantum number, and allowed electron states lie on tubes of constant magnetic quantum number m. The tubes extend to infinity but are truncated to show regions occupied by electrons. As the magnetic field increases, the tubes swell and the occupancy of the m = 4 level falls to zero as it passes through the Fermi level, etc.

the zero-field case, otherwise there would be a large diamagnetism (see Section V. B). As the field H increases, the orbital area S of each tube will increase according to Eq. (45), passing through the electrochemical potential η at a field H given by Eq. (46). For most metals the spacing between levels of quantum numbers m and m + 1 ($= \hbar\omega_c$) is relatively small, but in graphite, since the effective mass is very small, $\hbar\omega_c$ may become as large as the chemical potential. Assuming m* = 0.05m_0, then $\hbar\omega_c$ = 2.35 x 10^{-2} eV at 100 kG (10 teslas). Thus, quantum effects are expected to play a very important role, particularly at low temperatures (kT << $\hbar\omega_c$), where thermal broadening of the levels is unimportant.

B. Magnetic Energy Levels in Graphite for $\gamma_3 = 0$

The energy levels for the case $\gamma_3 = 0$ have been calculated independently by McClure [66] and Inoue [67] (see also Ref. [68]). Both authors derived the Hamiltonian by the method of Luttinger and Kohn [69], from which the secular equation could be obtained:

$$\left(\frac{3\gamma_0^2 a_0^2}{2} \frac{eH}{\hbar c}\right)^2 m(m+1) + \frac{3\gamma_0^2 a_0^2}{2} (m + \tfrac{1}{2}) \frac{eH}{\hbar c} \left[\frac{\epsilon - \epsilon_1^0}{(1 - \nu)^2} + \frac{\epsilon - \epsilon_2^0}{(1 + \nu)^2}\right] (\epsilon - \epsilon_3)$$

$$+ \frac{(\epsilon - \epsilon_3^0)^2 (\epsilon - \epsilon_1^0)(\epsilon - \epsilon_2^0)}{(1 - \nu^2)^2} = 0. \qquad (47)$$

This is a quartic equation in ϵ that does not factorize into two quadratic equations as in the case for H = 0 (Section III. B). Using approximations to solve for ϵ, the following semiclassical result is obtained for large m:

$$\epsilon_m^{\pm}(1) = \frac{\epsilon_1^0 + \epsilon_3^0}{2} \pm \left|\left(\frac{\epsilon_1^0 - \epsilon_3^0}{2}\right)^2 + \frac{3}{2}(1 - \nu)^2 a_0^2 \gamma_0^2 (m + \tfrac{1}{2}) \frac{eH}{\hbar c}\right|^{1/2}, \qquad (48)$$

$$\epsilon_m^{\pm}(2) = \frac{\epsilon_2^0 + \epsilon_3^0}{2} \pm \left|\left(\frac{\epsilon_2^0 - \epsilon_3^0}{2}\right)^2 + \frac{3}{2}(1 + \nu)^2 a_0^2 \gamma_0^2 (m + \tfrac{1}{2}) \frac{eH}{\hbar c}\right|^{1/2}. \qquad (49)$$

This interesting result shows that the same result is obtained as for the zero-field case [Eqs. (11) and (12)], with κ^2 replaced by 2(m + 1/2) (eH/$\hbar$c).

In the corners of the zone ($\cos \phi = 0$) the energy can be written down exactly:

$$\epsilon = \frac{\Delta}{2} \pm \left| (\frac{\Delta}{2})^2 + \frac{3\gamma_0^2 a_0^2}{2} (m, \ m + 1) \frac{eH}{\hbar c} \right|^{1/2} , \tag{50}$$

where $(m, \ m + 1) = m$ or $(m + 1)$.

For small values of the magnetic quantum number m, the levels are [66] approximately

$$\epsilon = \epsilon_3^0 + \frac{3\gamma_0^2 a_0^2}{2} \frac{eH}{\hbar c} \left[-\frac{1}{2} (m + \frac{1}{2}) (\omega_1 + \omega_2) \right.$$

$$\left. \pm \left| \left(\frac{m + \frac{1}{2}}{2}\right)^2 (\omega_1 - \omega_2)^2 + \frac{\omega_1 \omega_2}{4} \right|^{1/2} \right] \tag{51}$$

and

$$\epsilon = \epsilon_{1,2} + \frac{3\gamma_0^2 a_0^2}{2} \frac{eH}{\hbar c} (m + \frac{1}{2}) \omega_{1,2} , \tag{52}$$

where

$$\omega_1 = \frac{(1 - \nu)^2}{\epsilon_1^0 - \epsilon_3^0} ; \quad \omega_2 = \frac{(1 + \nu)^2}{\epsilon_2^0 - \epsilon_3^0} ; \quad \epsilon_3^0 \neq \epsilon_1^0 \text{ or } \epsilon_2^0 .$$

It is noted that for m = 0 the positive root in Eq. (51) gives a solution $\epsilon = \epsilon_3^0$, so that the zero-point energy $(\hbar \omega_c / 2)$ is missing. This result was derived for the two-dimensional band model by McClure [70].

Finally, for the special case $\epsilon_2^0 = \epsilon_3^0$ (i.e., $\phi = \phi_c$), the energy levels are given for small m by

$$\epsilon = \epsilon_3^0 \pm (1 - \nu) \left| \frac{3\gamma_0^2 a_0^2}{2} \frac{eH}{\hbar c} (m + \frac{1}{2}) \right|^{1/2} , \tag{53}$$

$$\epsilon = \epsilon_3^0 - \frac{m(m + 1)}{(m + 1/2)} \omega_1 \frac{3\gamma_0^2 a_0^2}{2} \frac{eH}{\hbar c} , \tag{54}$$

$$\epsilon = \epsilon_1^0 + (m + \frac{1}{2})\omega_1 \frac{3\gamma_0^2 a_0^2}{2} \frac{eH}{\hbar c} . \tag{55}$$

McClure noted that, in order to find the quantum number in terms of the energy, it is only necessary to solve a quadratic equation:

$$(m + \frac{1}{2})B = \frac{(\epsilon - \epsilon_3^0)}{2}\left[\frac{\epsilon - \epsilon_1^0}{(1 - \nu)^2} + \frac{\epsilon - \epsilon_2^0}{(1 + \nu)^2}\right]$$

$$\pm\left(\left\{\left[\frac{(\epsilon - \epsilon_3^0)}{2}\right]\left[\frac{\epsilon - \epsilon_1^0}{(1 - \nu)^2} - \frac{\epsilon - \epsilon_2^0}{(1 + \nu)^2}\right]\right\} + \frac{B^2}{4}\right)^{1/2}, \tag{56}$$

where $B = \dfrac{\gamma_0^2 a_0^2}{2}\dfrac{eH}{\hbar c}$.

By the use of this formula, the energy for the levels $m = 0, \ldots, 11$ has been established as a function of ϕ at 50 kG by M. S. Dresselhaus and Mavroides [65] (Fig. 14).

C. Magnetic Energy Levels in Graphite for Nonzero γ_3

Inoue [67], using a perturbation method, calculated the effect of γ_3 on the magnetic energy levels. The perturbation (δ) on the energy levels themselves was found to be small. At $\phi = 0$, $\delta \simeq (\gamma_3^2/\gamma_0^2)\,\gamma_1 \simeq 10^{-3}$ eV for $\epsilon_m^-(1)$ and $\epsilon_m^+(2)$, and is negligible for $\epsilon_m^+(1)$ and $\epsilon_m^-(2)$. The most important effect of the trigonal warping is that it allows electrons to be excited by incident radiation for harmonics of the type $\delta m = \pm(1+3N)$ (N an integer) [67], whereas the fundamental occurs for the selection rule $\delta m \pm 1$. These calculations were made for $\gamma_3 \leq 0.2$ eV.

Recent results of Schroeder et al. [45] lead to the conclusion that $\gamma_3 = 0.29 \pm 0.02$ eV (see Section VI. B), so that γ_3 may not be treated as a perturbation, as assumed by Inoue [67]. The 4 x 4 Hamiltonian used to derive the secular Eq. (47) must then be replaced by one of infinite extent in the exact case. Schroeder et al. [45] have shown that a 36 x 36 truncated Hamiltonian gave correct results for the energy of the Landau levels up to $m = 10$, indicating that terms up to tenth order in γ_3 needed to be included to achieve convergence. Their results indicate that γ_3 has a strong influence on the energy of the levels as well as on the selection rules.

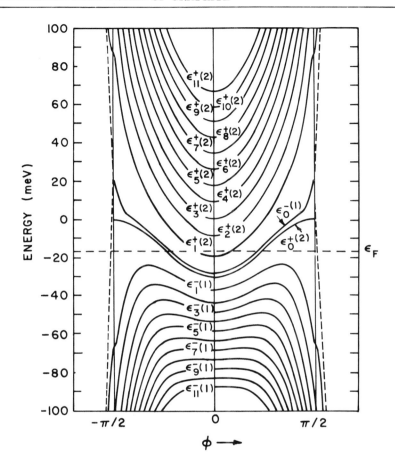

FIG. 14. Contours of constant energy at 50 kG (5 teslas) for the four π bands in graphite. The labeling of states is the same as that used in the text. Each contour describes the variation in energy with the reciprocal lattice vector k_z along tubes of constant magnetic quantum number. The $\epsilon_{m=1}^{-}(1)$ level (hole) has passed through the Fermi level, but the $\epsilon_{m=1}^{+}(2)$ state (electron) does not pass through until about 70 kG. The level $\epsilon_0^{+}(2)$ is independent of field and marks the lower edge of the conduction band, and $\epsilon_0^{-}(1)$ the upper edge of the valence band. Spin-orbit-splitting effects are not included in the diagram. Adapted from M.S. Dresselhaus and Mavroides [65].

V. DENSITY OF STATES AND ELECTROCHEMICAL POTENTIAL

A. Density of States in Zero Magnetic Field

The energy density of states can be calculated once the dispersion relationship is known. Since for unit volume of solid reciprocal space contains one electron of each spin value in a volume $1/8\pi^3$, the problem is reduced to finding volumes between constant-energy surfaces.

For the simplest model of cylindrical bands the energy density of states $n(\epsilon)$ is independent of energy. Allowing two complete cylinders of height $2\pi/C_0$ in the Brillouin zone, the density for both spin directions is

$$n(\epsilon)\, d\epsilon = \frac{2m}{\pi\hbar^2 C_0}\, d\epsilon, \tag{57}$$

which compares with the more normal dependence on the square root of the energy for a single spheroidal band,

$$n(\epsilon)\, d\epsilon = \frac{1}{2\pi^2}\, \frac{\sqrt{8m_{11}^2 m_{33}}}{\hbar^3} \epsilon^{1/2}\, d\epsilon. \tag{58}$$

For the four-parameter band model we can proceed as follows: Considering the electron surface ϵ that encloses carriers for $-\theta < \phi < \theta$, where θ is the value of ϕ for which $\epsilon = \epsilon_3^0$,

$$\cos\,\theta = \sqrt{\frac{\epsilon}{2\gamma_2}}. \tag{59}$$

The total number of carriers $N(\epsilon)$ with energy less than ϵ for both spin directions in two volumes of revolution is

$$N(\epsilon) = \frac{4}{8\pi^3} \int_{|\phi|<\theta} \pi\kappa^2\, dk_z = \frac{1}{\pi^2 C_0} \int_{-\theta}^{\theta} \kappa^2\, d\phi. \tag{60}$$

Therefore the density of states is

$$n(\epsilon) = \frac{dN(\epsilon)}{d\epsilon} = \frac{1}{\pi^2 C_0}\, \frac{d}{d\epsilon} \int \kappa^2\, d\phi = \frac{1}{\pi^2 C_0} \int 2\kappa \frac{d\kappa}{d\epsilon}\, d\phi, \tag{61}$$

so that for the parabolic model with $\Delta = 0$:

$$n^{el}(\epsilon) = \frac{16}{3\pi^2}\, \frac{\gamma_1}{\gamma_0^2 C_0 a_0^2}\, \sin\,\theta. \tag{62}$$

where

$$\sin \theta = \left(1 - \frac{\epsilon}{2\gamma_2}\right)^{1/2} \quad \text{if} \quad 2\gamma_2 < \epsilon < 0,$$

$$\sin \theta = 1 \quad \text{if} \quad \epsilon > 0. \tag{63}$$

The change in the function $\sin \theta$, and hence in $n(\epsilon)$, at $\epsilon = 0$ comes about when the tips of the constant-energy surface touch the hexagonal face of the Brillouin zone. It is interesting to note that for electron energies greater than $\epsilon = 0$ (i.e., for energies greater than $|2\gamma_2|$ above the bottom of the conduction band) the density of states $n(\epsilon)$ is independent of energy, as for the cylindrical band model, Eq. (57).

A relationship similar to Eq. (62) holds for the hole states:

$$n^h(\epsilon) = \frac{16}{3\pi^2} \frac{\gamma_1}{\gamma_0^2 C_0 a_0^2} (1 - \sin \theta). \tag{64}$$

These results, first obtained by Nozieres [71, 72] in a form compatible with positive γ_2, are sketched in Fig. 15.

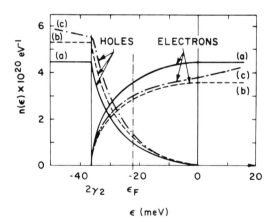

FIG. 15. The density of states for carriers in graphite near the region of band overlap: (a) four-parameter parabolic model; (b) four-parameter parabolic model with γ_4 included; (c) four-parameter hyperbolic model with γ_4 included. (γ_2 is taken as negative in this figure.)

The result obtained in a similar way for the four-parameter hyperbolic model is

$$n^{e\ell}(\epsilon) = \frac{16\gamma_1}{3\pi^2\gamma_0^2 C_0 a_0^2}\left[\sin\theta + \frac{(\epsilon - \Delta - 2\gamma_2)}{2\gamma_1}\theta - \frac{\gamma_2}{2\gamma_1}\sin\theta\cos\theta\right], \qquad (65)$$

where $|\theta| \leq \pi/2$. A similar result can be obtained for the hole states. Thus the density of states varies linearly with energy outside the bounds 0, $2\gamma_2$ (Fig. 15). Since the value of $(\epsilon - \Delta - 2\gamma_2)$ is small compared with values of $2\gamma_1$, however, this variation is quite small. Finally, the parameter $\nu = (2\gamma_4\cos\theta)/\gamma_0$ reduces the electron density of states by a factor on the order of $(1 + \nu_0)^{-2}$, where $\nu_0 = 2\gamma_4/\gamma_0$ and increases the density of states for holes by an approximate factor $(1 - \nu_0)^{-2}$ (Fig. 15).

B. Electrochemical Potential and Fermi Level

At a temperature T, the number of free electrons in the bands can be calculated from

$$n = \int n(\epsilon)\, f_0(\epsilon)\, d\epsilon, \qquad (66)$$

where $f_0 = [\exp(\epsilon - \eta)/kT + 1]^{-1}$ is the Fermi-Dirac distribution function, η is the electrochemical potential of the carriers, and where the integration is taken over all energy states. In this chapter the use of the term "Fermi energy" is reserved for the very-low-temperature region ($kT \ll \eta$) applicable to extreme degeneracy.

Fermi-Dirac integrals $Fn = \int_0^\infty x^n f_0(x)\, dx$, where x is the reduced energy (ϵ/kT), occur frequently in descriptions of free-carrier properties and can be integrated by parts to give

$$Fn = \int_0^\infty x^n\, f_0(x)\ dx = \frac{1}{n+1}\int x^{n+1}\frac{\partial f_0}{\partial x}\ dx \qquad (67)$$

provided that $n > -1$ [52]. Tabulated values of the Fermi-Dirac integral Fn for half-integral values of n are given by McDougall and Stoner [73] and Beer et al. [74] for integral values of n by Rhodes [75], but with modern high-speed computers it is easier to evaluate the integrals directly.

By using Eq. (66) the Fermi energy and shift of the electrochemical potential with temperature can be obtained from the condition that the number of filled states in the conduction band must equal the number of empty states in the valence band, allowing for differences arising from impurities. For the four-parameter parabolic model the Fermi energy is given by

$$\epsilon_F = \frac{4\gamma_2}{3},$$ (68)

providing that $\Delta = 0$, so that the number of electrons and holes is

$$p = n = \frac{64\gamma_1\gamma_2}{9\sqrt{3}\pi^2\gamma_0^2 a_0^2 C_0}$$ (69)

Arkhipov et al. [76] have shown that the electrochemical potential for this model does not depend on temperature. This may be seen immediately from an expression given by Kechin [25] in an extension of the model for which $\gamma_4 \neq 0$:

$$\eta = \frac{4}{3}\gamma_2 + \frac{8\gamma_4 kT}{\gamma_0}\int_0^{\frac{\pi}{2}} \log\left| 2\cosh\left(\frac{\epsilon_3^0 - \eta}{kT}\right) \right| \cos^2\phi \, d\phi.$$ (70)

This result is obtained from the formulas for n and p, using effective masses depending on the parameter γ_4 [see Table 2]

$$n = \frac{4}{\pi^3\hbar^2}\int_0^{\pi/2} m_{el}^*(\phi)d\phi \int_{\epsilon_3^0}^{\infty} f_0 d\epsilon$$

$$= p = \frac{4}{\pi^3\hbar^2}\int_0^{\pi/2} m_h^*(\phi)d\phi \int_0^{\epsilon_3^0} (1-f_0)d\epsilon \, .$$ (71)

A variation in the electrochemical potential is expected whenever there are differences in the densities of states of holes and electrons. Figure 16 sketches the variation in the electrochemical potential with temperature for Eq. (70), and compares the result obtained by Ono and Sugihara [77] for the four-parameter hyperbolic model.

A discussion of the number of free carriers is given in Section VII.G.

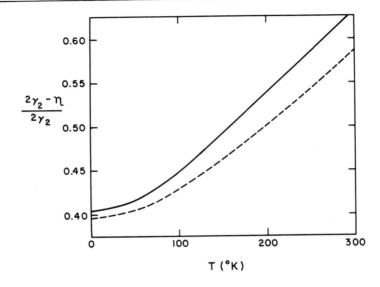

FIG. 16. Variation of electrochemical potential with temperature obtained by Kechin [25] for the four-parameter parabolic band model including the effect of γ_4 (broken curve), and by Ono and Sugihara [77] for the four-parameter hyperbolic band model including γ_4 (solid curve). The quantity $(2\gamma_2-\eta)/2\gamma_2$ represents the relative position of the electrochemical potential with respect to the bottom of the conduction band.

C. Degeneracy of Magnetic Energy Levels and Shift of Fermi Level with Magnetic Field

When a magnetic field is applied, the wavevector $\underline{k}$ is no longer a good quantum number. Allowed values in reciprocal space are no longer represented by a lattice of points, each occupying a volume $1/8\pi^3$, but now lie on the tubes of constant magnetic quantum number. For each quantum number m the number of states per interval $d\phi = C_0/2 \cdot dk_z$ [66] is:

$$= \frac{2eH}{\hbar c C_0 \pi^2} \, d\phi, \tag{72}$$

which allows the density of states and the Fermi level to be computed in the field H. The shift of Fermi energy with the magnetic field predicted by Sugihara and Ono [78] is sketched in Fig. 17. Kinks in the curves correspond to magnetic field values at which magnetic energy levels pass through the Fermi surface, so that the condition for coincidence is drastically modified by the shift in ϵ_F.

Woollam [79] has compared the field values for crossings with experimental data, using the Shubnikov-De Haas effect (see Section VI. F). Agreement is poor, notably for the m = 1 electron level, which crosses the Fermi

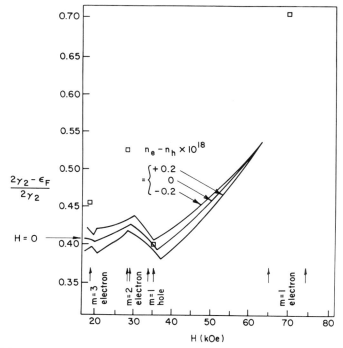

FIG. 17. Variation of the Fermi energy with magnetic field at very low temperatures. The solid lines represent a calculation of Sugihara and Ono [78] for intrinsic graphite and graphite with small concentrations of acceptor and donor defects. Vertical arrows represent experimental values of coincidences of magnetic energy levels and the Fermi level [79]. Values of $\epsilon_F(H)$ calculated from the experimental data are also indicated. [□] Poor agreement between experiment and theory is probably due to the choice of band parameters used by Sugihara and Ono.

level at a much higher value (~70 kG) than that predicted by Sugihara and Ono [78] (~38 kG). By fitting Eq. (47) to experimental data for the magnetic field values, the Fermi energy can be computed from the condition

$$\epsilon_F = (m + \xi)h\omega_c,$$ (73)

where ξ is the phase and is equal to 1/2 for free electrons.

The observed shift (Fig. 16) is therefore to values of energy higher than predicted, relative to the bottom of the conduction band.

It is possible to resolve the separation in magnetic energy levels due to spin-orbit interaction for the m = 1 and m = 2 electron levels and m = 2 hole level [79]. This effect was not considered by Sugihara and Ono [78] and is not reflected in Fig. 17. A further discussion of this effect is given in Section VI.G.

The above calculations were made for the case of complete degeneracy (T = 0^OK). At higher temperatures the carriers occupy states with energy within about $\pm$ kT of the degenerate magnetic level. At 300^OK (kT $\simeq$ 24 meV) it is clear that quantum effects will be unimportant and the shift of Fermi level with magnetic field will be smoothed out. However, at 77^OK (kT $\simeq$ 6 meV) it should be possible to observe quantum effects in high fields, (H $\geq$ 20 kG) particularly in effects that depend sensitively on differences in carrier properties (see Section VII.H).

VI. EXPERIMENTAL DETERMINATION OF BAND PARAMETERS IN GRAPHITE

A. Cyclotron Resonance

In the simplest case of a free-electron solid, incident radiation may couple effectively with the electrons when the photon energy coincides with the energy difference between magnetic energy levels, subject to the selection rule $\delta m = \pm 1$. The dominant response of the system comes from electron orbits for which the energy difference between magnetic energy levels is stationary with respect to the wavevector parallel to the applied field. Since holes and electrons rotate in opposite senses in a magnetic field, resonance absorption of holes and electrons for a given magnetic field direction will occur for different polarizations of incident energy or alternatively, if the polarization is held constant, for different directions of the magnetic field (see, for example, Ref. [57]). In principle it is therefore possible to distinguish holes and electrons by using this effect.

Cyclotron-resonance effects were first observed by Galt, Yager, and Dail [80] on purified natural crystals of graphite. In their experiment the incident microwave radiation was parallel to the applied magnetic field and to the c-axis of the sample. Circularly polarized radiation was used in the experiments, with the ratio of the polarizations in the two directions maintained at 15:1. The symmetrical power-absorption curve for different magnetic field directions (Fig. 18a) indicates an equal number of electrons and holes. Plots of the power derivative dP/dH obtained by a field-modulation technique show sharp resonances on both the electron (-H) and hole (+H) sides of the curves (Fig. 18b).

Following a preliminary explanation of the data with discrete hole and electron masses [81], Nozieres [71, 72] used the four-parameter parabolic model with a distribution of effective masses. He concluded that the sharp peaks in dP/dH originated from minority carriers near the crossover of ϵ_F and ϵ_3^0 where trigonal warping produces the satellite pieces. Recognizing that trigonal warping modifies the selection rule so that transitions of the type $\delta m = \pm(1 + 3N)$ are allowed, where N is an arbitrary integer, he was able to explain the data if <u>holes</u> occupy states near point K. Thus γ_2 was taken as <u>positive</u>. This interpretation was largely responsible for the adoption of the heavy-hole model until quite recently.

Inoue [67] calculated the magnetic energy levels (Sections IV. B and IV. C) and theoretically calculated the resonance conditions for cyclotron resonance. He concluded that if γ_2 were positive, resonance effects could arise from holes at point K and electrons at the crossover of the levels ϵ_2^0 and ϵ_3^0. With <u>negative</u> γ_2, this would read "electrons at point K and holes at the crossover of ϵ_1^0 and ϵ_3^0." In either case the crossover has been shown not to occur, since the parameter Δ is positive. Inoue also obtained the selection rule

$$\delta m = \pm (1 + 3N), \tag{74}$$

where N is an arbitrary integer, and concluded that the elimination of either hole or electron signals by using polarized radiation was not possible because of the complicated band structure. Both hole and electron signals are mixed.

Inoue [67] discussed the band parameters obtainable from the cyclotron-resonance experiments by assuming that signals from both types of carrier could be observed. The experiments have recently been repeated with both

(a)

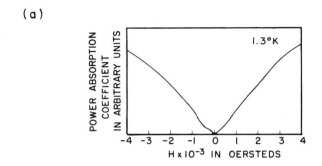

(b)

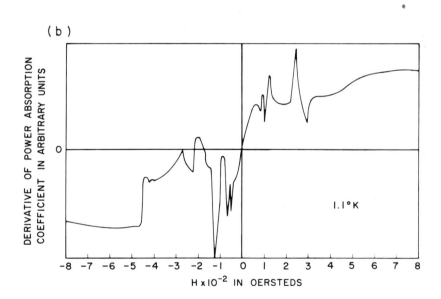

FIG. 18. Plots of (a) the power-absorption coefficient against magnetic field at 1.3°K with circularly polarized radiation at 24,000 MHz and magnetic field normal to the basal planes, and (b) of the derivative of the curve in (a), obtained with a field-modulation technique. Reprinted from Ref. [80] by courtesy of the American Institute of Physics.

natural and synthetic crystals [82] and with more sensitive detection apparatus. Since all the peaks shifted in the same manner toward higher field
values as the field was tipped away from the c axis, Williamson et al. [82]
concluded that all of the observed resonances could be assigned to the majority carrier at point K. Using the condition for resonance,

$$\hbar\omega = (3N + 1)\hbar\omega_c, \tag{75}$$

and the approximate expression for carriers at point K [see Eqs. (49) and
(20)],

$$\omega_c = \frac{eH}{m^*c} = \frac{3\gamma_0^2 \, a_0^2 (1 + \nu_0)^2}{4\nu_2 \, \hbar^2} \frac{eH}{c} \tag{76}$$

a straight-line plot of (3N + 1) versus 1/H yielded a value of m* = 0.061 ±
0.002 m_0 (Fig. 19).

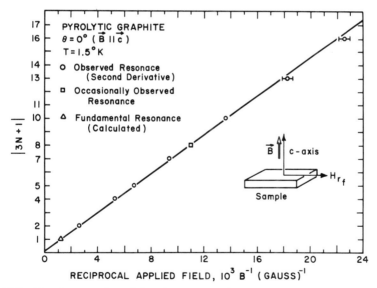

FIG. 19. Integer plot of the observed cyclotron resonances for normally
incident radiation. The straight line represents the condition $(3N + 1)\hbar\omega_c =$
$\hbar\omega$. The data are consistent with an effective mass for electrons at $\phi = 0$
(and with energy close to the Fermi energy) m* = $(0.061 \pm 0.002)m_0$. Reprinted from Ref. [82] by courtesy of the American Institute of Physics.

A resonance condition at point H was examined. Since the constant-energy surfaces are circular in the basal plane, the minority-carrier pocket there should not produce harmonics. However, it was concluded [82] that the signal strength would be too low, since the number of carriers associated with the pocket is low and since the resonance would be hidden by the harmonics of the majority carriers.

Assuming that the resonances can be attributed to the majority carriers at point K, Schroeder and co-workers [20] reexamined the data of Galt et al. [80]. By identifying the resonance condition at the center of the steep portion of the absorption derivative curve (Nozieres [71] took the maxima of the first derivative while Williamson et al. [82] used the minima of the second derivative, which should correspond closely with the condition used by Schroeder et al. [20]) and measuring the strength of the resonance from the height of the steep section of the curve, they were able to identify the peaks on both sides of the absorption curve as arising from electrons at point K with mass $0.061 \pm 0.002 \ m_0$. In view of the complexity of the analysis, the assignment of electrons at point K is not definitive, requiring the sharper tool afforded by magnetoreflection experiments.

B. Magnetoreflection

Oscillations in the optical reflectivity from basal surfaces of graphite are observed with a magnetic field parallel to the c-axis. The origin of the effect is closely analogous to that of cyclotron resonance, except that for higher photon energies interband transitions predominate with the selection rule $\delta m = \pm 1$ for the fundamental (Figs. 20 and 21).

Earlier work [44, 65] with unmonochromatized infrared radiation identified two series of peaks (Fig. 22). The K series corresponds with interband transitions at point K in the Brillouin zone, with electrons excited from filled magnetic energy levels in the valence band to unfilled levels above the Fermi level in the conduction band. For this series, harmonics corresponding to the selection rule $\delta m = \pm(3N + 1)$ have been observed in the higher frequency end of the spectrum. The second series (H series) corresponds to interband transitions between degenerate ϵ_1 or ϵ_2 levels to degenerate ϵ_3 levels, with the selection rule $\delta m = \pm 1$ only, as predicted, but not observed, for cyclotron resonance. Since the energy levels at H are given by Eq. (50), the experiment can give the parameters Δ and γ_0 from a plot of the incident

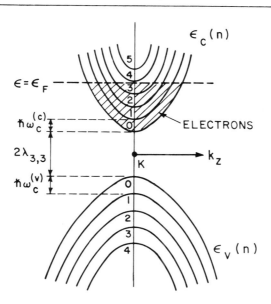

FIG. 20. Magnetic energy levels for the carriers near point K. The difference in the energy separation ($\hbar\omega_c$) for the valence and conduction bands can be used to estimate γ_4. Interband transitions, subject to the selection rule $\delta m = \pm(1+3N)$, can occur only when an electron is excited from an occupied state in one band to an unoccupied state in the other band. Adapted from Ref. [83].

energy $\hbar\omega$ against the index of the transition (Fig. 23). At point K, with solutions for ϵ_m including effect of γ_3, it is possible to obtain γ_1, γ_3, γ_4. The latest value for γ_3 obtained ($\gamma_3 = 0.3$ eV) [45] indicates that the electron surfaces are much more strongly warped than De Haas-Van Alphen results suggest [84] and of such magnitude that perturbation treatments of the effect of γ_3 on the bands are not adequate.

In the later studies quoted [20, 45] a monochromatic laser beam was used, giving higher resolution. The selection rules for the different senses of polarization with respect to the magnetic field direction [$m_{(v)} \rightarrow (m+1)_{(c)}$ for one sense; $m_{(v)} \rightarrow (m-1)_{(c)}$ for the other sense] allowed a definitive sign to be attached to γ_4 (positive). The authors then stated that, in order to maintain consistency with De Haas-Van Alphen and Shubnikov-De Haas

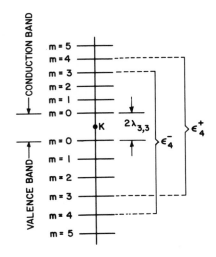

$$\epsilon_4^- = \epsilon_c(3) - \epsilon_v(4)$$
$$\epsilon_4^+ = \epsilon_c(4) - \epsilon_v(3)$$
$$\left.\right\} \delta m = \pm 1$$

FIG. 21. Allowed interband transitions at point K, $\epsilon_n^- = \epsilon_c(n - 1) - \epsilon_v(n)$ and $\epsilon_n^+ = \epsilon_c(n) - \epsilon_v(n - 1)$. Since the effective masses in the two bands are different, $\epsilon_n^+ \neq \epsilon_n^-$. Transitions ϵ_4^- and ϵ_4^+ are shown corresponding to different senses of polarization. This allows the valence and conduction bands to be calculated, as well as the spin-orbit parameter $\lambda_{3,3}$. The particular case drawn is for $m_{(v)} < m_{(c)}$ $(\hbar\omega_v > \hbar\omega_c)$, which is opposite to the actual situation in graphite.

---

data, it was necessary for γ_2 to be negative and γ_4 positive. Since this statement has not been amplified and its meaning is not immediately obvious, it is dealt with more fully here.

The magnetoreflection experiment allows the two transitions

$$\epsilon_n^- = \epsilon_c(n - 1) - \epsilon_v(n) = \hbar\omega^-, \tag{77a}$$

$$\epsilon_n^+ = \epsilon_c(n) - \epsilon_v(n - 1) = \hbar\omega^+, \tag{77b}$$

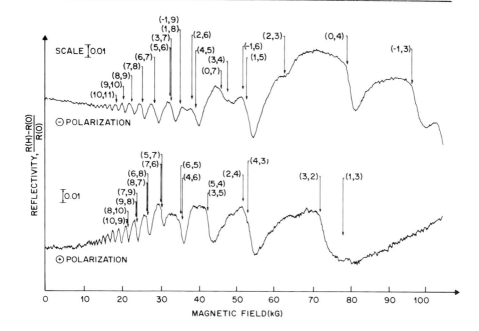

FIG. 22. Experimental recorder traces obtained from a laser magneto-reflection experiment (pyrolytic graphite, $\hbar\omega = 0.069336$ eV). Resonance-field values predicted from the Slonczewski-Weiss [12] model are indicated by arrows. For each transition the Landau level indices for the valence and conduction levels are indicated by the first and second numbers in parentheses, respectively. The field direction and geometry are such that electron cyclotron resonance would occur for the traces labeled $\ominus$ polarization and hole resonances for those labeled $\oplus$ polarization. From Ref. [45].

to be distinguished from each other from the sense of polarization relative to the magnetic field. Once the indexing of the peaks (Fig. 22) is accomplished, the effective mass for the conduction and valence bands can be computed from Eqs. (77a) and (77b), and the relationships given in Section IV.A for magnetic energy levels (see Fig. 21). The actual values are 0.10 m_0 for the valence band and 0.056 m_0 for the conduction band [18].

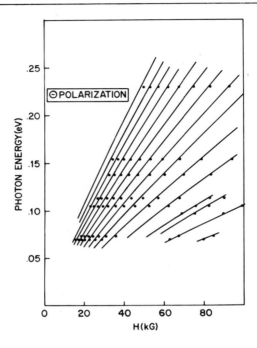

FIG. 23. A summary of the magnetoreflection results for the K series of transitions. Each curve corresponds to an interband transition. The points are experimental data, whereas the solid lines represent the predictions of the Slonczewski-Weiss band model. From Ref. [45].

Irrespective of the sign of γ_2, the valence and conduction bands are to be associated with the $\epsilon_m^-(1)$ and $\epsilon_m^+(2)$ bands, respectively. Neglecting the nonparabolicity of the bands, the effective masses at K can therefore be written as

$$m^*_{c_{11}} = \frac{\hbar^2 (2\gamma_1 + 2\gamma_2 - \Delta - 2\gamma_5)}{2\eta_0^2 (1 + \nu_0)^2}, \qquad (78a)$$

$$m^*_{v_{11}} = \frac{\hbar^2 (\Delta + 2\gamma_1 - 2\gamma_2 + 2\gamma_5)}{2\eta_0^2 (1 - \nu_0)^2}. \qquad (78b)$$

Assuming for the moment that $\nu_0 = 2\gamma_4/\gamma_0 = 0$, it can be seen that differences in the effective masses can be explained on the basis of differences in the numerators of Eqs. (78). In particular a large positive value of Δ or a large positive value of γ_5 could produce the required differences. However, other evidence suggests that both of these parameters are small. In order to explain the effective-mass differences consistently it is required that γ_4 be greater than zero.

These data taken alone still do not assign a definitive polarity to γ_2. However, De Haas-Van Alphen measurements give the effective mass of the carrier at point K as $0.056m_0$, and hence this carrier must be an <u>electron</u>, and therefore, γ_2 <u>negative</u>.

C. De Haas-Van Alphen Oscillations

The oscillatory component of the magnetic susceptibility comes about from the quantization of the energy levels in a magnetic field, discussed in Section IV. As the field increases, the energy of the magnetic levels changes until eventually the Fermi energy is surpassed. As each magnetic level passes through the Fermi surface, the free energy of the electron assembly changes abruptly, so that a corresponding change is observed in the susceptibility (De Haas-Van Alphen effect) and other properties, such as the magnetoresistance and Hall effect (Shubnikov-De Haas effect). The oscillations are periodic in the reciprocal of the magnetic field, with period $\Delta(1/H)$ related to the cross-sectional area of the Fermi surface perpendicular to the field S, by

$$\Delta(\frac{1}{H}) = \frac{2\pi e}{\hbar c S} . \tag{79}$$

The effect is only seen for maximum or minimum cross sections of the Fermi surface and at low temperatures (T $\leq$ 4°K), where the thermal broadening of the magnetic energy levels is much less than the level separation.

$$kT \ll \hbar\omega_c. \tag{80}$$

Earlier workers [85-89] observed two periods [$\Delta(1/H) \simeq 1.5 \times 10^{-5}$ and 2.1×10^{-5} gauss^{-1}] (Table 3). The shorter period (larger extremal area) may be identified in the Slonczewski-Weiss [12] model with the majority carriers at K, and the longer period with the majority carriers at the extremal section of the two "wing" surfaces, with cross sections given

TABLE 3

Data Obtained from De Haas-Van Alphen and Shubnikov-De Haas Experiments

Type of crystal	Type of measurement	Period of oscillation (10^5 gauss^{-1})			Dingle temperature (°K)			Refs.
		Majority hole	Majority electron	Minority carriers	Majority hole	Majority electron	Minority carriers	
Natural	Susceptibility	2.20	1.65		1.5 / 0.5[a]	0.9 ± 0.6 / 0.6[a]		85
Natural	Susceptibility / Magnetoresistance / Hall coefficient	2.15 / 2.15 / 2.15	1.61		0.71[b]	0.64[b]		86
Natural	Magnetoresistance / Hall coefficient	2.12 ± 0.01 / 2.10 ± 0.001	1.59 ± 0.01 / 1.57 ± 0.01		3.6[b]	3.4[b]		87, 88
Natural	Magnetoresistance	2.07 ± 0.04	1.51 ± 0.03			0.56		84
Natural	Susceptibility	2.19	1.60	13.5 ± 3, ~25	0.80			90
Natural	Susceptibility	2.08 ± 0.02	1.51 ± 0.03	13.5 ± 0.1			1.5 ± 0.1	5, 94
Synthetic	Susceptibility	2.08 ± 0.02	1.51 ± 0.03	22.4 ± 0.3				5, 94
Synthetic	Magnetoresistance		1.51					91
Natural	Susceptibility	2.09 ± 0.002	1.50 ± 0.003	13.3 ± 0.9				92
Natural	Magnetoresistance	2.252	1.628		1.70	0.62		22
Synthetic	Magnetoresistance	2.188	1.613		4.21	2.58		22
Synthetic	Hall coefficient and thermopower			12.0 ± 1				93

[a] Values computed by Soule, McClure, and Smith [84].

[b] Values computed by Soule, McClure, and Smith [84] for measurements on same crystal showing much lower value obtained for galvanomagnetic oscillations than for susceptibility effect.

by Eqs. (17) and (18), respectively. It is not possible to identify the carrier sign from the low-field oscillations. In a later paper, Soule [90] reported a new oscillation with a very long period [$\Delta(1/H) \approx 13.5 \times 10^{-5}$ gauss^{-1}] with a very small Fermi surface area. From measurements on both natural and synthetic crystals, Williamson, Foner, and Dresselhaus [5, 94] confirmed the new oscillation and suggested that it arose from the minority carriers near point H (see following section).

A most thorough study of the magnetoresistance oscillation due to the majority carriers has been made by Soule, McClure, and Smith [84]. The experimental oscillations were fitted by using a least-squares technique to a generalized Landau formula:

$$G_i(H) = H^{n_i} \sum_{r=1}^{\infty} W_i(r) \frac{u_i}{\sinh u_i} \exp\left(-\frac{r u_i \Delta T_i}{T}\right)$$

$$\times \cos\left|\frac{2\pi r}{\Delta_i(1/H)H} - \psi_i(r)\right|. \tag{81}$$

In this expression for the carrier i, $u_i = 2\pi^2 kT/\Delta E_i$, where ΔE_i is the spacing between levels ($\hbar\omega_i$); $W_i(r)$ is the amplitude of the rth harmonic of the carrier i; and $k\,\Delta T_i$ is the energy associated with the collision broadening of the levels. The Dingle temperature ΔT_i is given by

$$\Delta T_i = \frac{\hbar}{\pi k \tau_i}, \tag{82}$$

where τ_i is the time between collisions [95]. In Eq. (81), $W_i(r)$, the exponent n_i, and the phase factor $\psi_i(r)$ are allowed arbitrary values. The expression is only valid for magnetic fields such that the number of levels below the Fermi surface is large.

The fitting of experimental data to Eq. (81) is difficult since the periods of the two majority oscillations differ by a factor of only about 1.5. At low fields the fitting is made simpler since the oscillations from the heavier carrier (electrons at point K) are damped out almost entirely compared with those from the lighter carrier. The extremal orbits associated with these carriers are indicated in Fig. 24a and comparison of experimental data is made in Table 3.

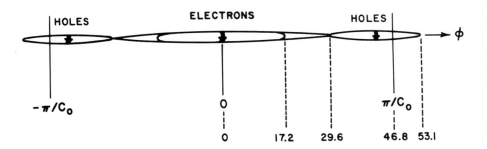

FIG. 24. The Fermi surface of graphite deduced from measurements of
the Shubnikov-De Haas periods [84]. Extremal orbits are indicated for a
magnetic field oriented parallel to the c-axis. Dimensions of the surfaces
are indicated in reciprocal units of 10^{+6} cm^{-1}. The anisotropies of the electron and hole surfaces are 17.3:1 and 12.1:1, respectively. The shorter
period must be identified with the extremal area on the plane $\phi = 0$ for the
number of electrons and holes to be equal.

Measurements were also made of the dependence of the period of the
oscillations $\Delta(1/H)$ on θ, the angle between the magnetic field and the c-axis
(Fig. 25). Measurements for the electron surface could only be made for θ
up to 85°, but the hole period could be followed all the way from $\theta = 0$ to 90°.
The results were consistent with a closed hole surface, similar to an ellipsoid (Fig. 24) with an axial ratio of 12.1:1. In order to accommodate an
equal number of electrons and holes it was necessary to fit conical tips to
the electron body, with an overall axial ratio of 17.3:1 (Fig. 24).

It is emphasized, however, that since the constant-energy surfaces are
highly elongated, the alignment of field perpendicular to the c-axis must be
effected very carefully. Furthermore, small misalignments of the c-axis
in differing parts of the specimens will invalidate conclusions as to whether
the surfaces are open or closed. For this reason no conclusions can be
reached with pyrolytic graphites, for which the c-axis misorientation is at
least 0.5° even in the best specimens [97]. Results on natural crystals
must be taken with some reservations until a detailed study of the crystal
misorientation is reported.

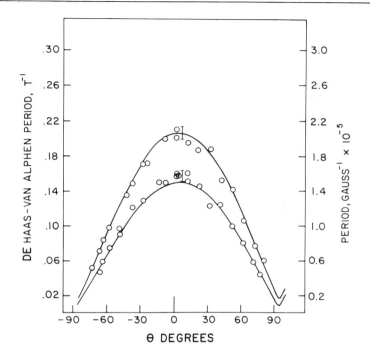

FIG. 25. Dependence of the majority-carrier periods of oscillation on θ, the angle between the magnetic field and the c axis. Data obtained by Soule et al. [84] for natural single crystals (curves) and those obtained by Woollam [96] for synthetic, highly ordered pyrolytic (PG3) crystals (open circles) are compared. From Ref. [96].

From the periods obtained from the fitting for both carriers the extremal cross-sectional areas can be determined, and from them the effective masses at the extremum (Table 3). The effective masses could also be obtained from the dependence on temperature (1-4°K) of the quantity sinh u_i. Similarly the dependence of the effective mass on the angle θ could be determined by both methods. Good agreement between the results of both methods is obtained, but the "temperature-dependent effective mass" is more reliable.

The results of Williamson et al. [5] on pyrolytic and natural graphite showed that the periods for the majority carriers in both specimens were the same within 2% (Table 3). This is an encouraging result, suggesting that the bands are essentially the same in these materials. Some differences were observed in the angular dependence of the periods (Fig. 25) and on the minority-carrier behavior, to be discussed in the next section, whereas Woollam [96] found that hole and electron cross-sectional areas in pyrolytic graphites agreed within experimental error with Soule's results on natural crystals out to 70°.

The low-field de Haas-Van Alphen oscillations themselves do not allow identification of the carrier signs to be made. Soule [90] has attempted to modify the Fermi level in a controlled way by using boron as a substitutional acceptor. On doping, the Fermi level should be depressed, so that, if γ_2 is negative, the electron extremal area should be decreased and the hole area increased. It follows that the de Haaas-Van Alphen periods should approach each other for small doping concentrations until a level is reached corresponding to equal periods. Further doping should then result in a divergence of the periods (Fig. 26). This prediction is quite different from that for positive γ_2, where the electron and hole extremal areas would again be reduced and increased respectively, corresponding to a monotonic divergence of the de Haas-van Alphen periods.

Soule's results [90], in which the scatter of points is high, indicated the periods to be monotonically divergent, corresponding to <u>positive</u> γ_2. McClure [17] has tentatively suggested that the levels of donor doping could have been larger than stated, so that the region of period convergence (low doping level) was missed entirely, but Soule [98] has pointed out that the same specimens gave a smooth variation of susceptibility with the stated doping levels (see Section VI.E).

This is the only experiment performed so far that gives a definitive value for the sign of γ_2 and predicts a positive value. A satisfactory explanation of the result will probably await further experiments.

More recently, Cooper et al. [22] have measured the periods of Shubnikov-de Haas oscillations for a smaller range of acceptor concentrations introduced by neutron irradiation. Their results show that the periods <u>converge</u> in the low concentration region, corresponding to a <u>negative</u> value of γ_2 (Fig. 26, bottom).

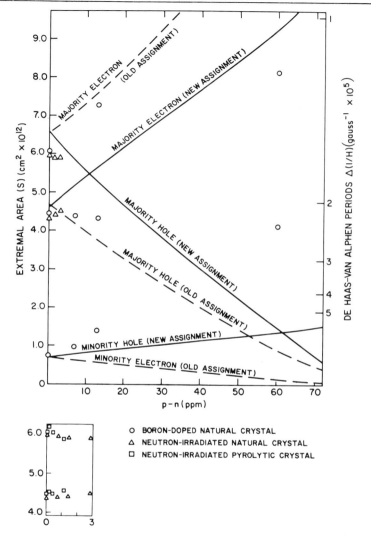

FIG. 26. Top: the predicted variation in the extremal areas of the
Fermi surfaces for holes and electrons in graphite as a function of acceptor
concentration. If γ_2 is negative, the areas initially converge, then cross
and diverge. If γ_2 is positive, the periods monotonically diverge. Bottom:
the limit of small acceptor concentration, showing the convergence of the
extremal areas obtained in neutron-irradiated graphite. After McClure
[18]. Data for boron-doped natural crystal (O) from Soule [90]; data for
neutron-irradiated graphite (Δ and $\square$) from Cooper et al. [22].

Attempts have been made to fit the experimental data for the Shubnikov-de Haas effect to a quantum theory of the conductivity developed by Adams and Holstein [99]. Using a δ-function for the scattering potential and an ellipsoidal model for the energy bands, Soule et al. [84] were able to fit the data satisfactorily except for a monotonic drift with magnetic field. They expressed their result as

$$\sigma_{xx} = \sigma_0 + \sigma_{c\ell} + \sigma_{osc}, \tag{83}$$

were $\sigma_{c\ell}$ is the classical high-field conductivity, σ_{osc} describes the Shubnikov-de Haas Oscillations, and σ_0 is the extra term required to fit the midline drift.

A more satisfactory calculation by Sugihara and Ono [78] used (a) a scattering potential with a range of 6 Å [corresponding to the screening length for ionized-impurity scattering (see Section VII. C)], (b) the Slonczewski-Weiss [12] band structure, and (c) allowed for the shift of electrochemical potential with magnetic field. The experimental curves could then be fitted without recourse to terms other than $\sigma_{c\ell}$ and σ_{osc}.

Two independent measurements have been made of the effect of hydrostatic pressure on the majority-carrier periods. The results of Anderson and co-workers [92] were probably made under conditions of more nearly hydrostatic pressure than those of Itskevich and Fisher [91], who only reported values for the carriers near point K. Results shown in Table 4 give reasonable agreement between the values. Both sets of authors analyzed their data assuming that the rate of change of γ_1 and γ_2 could be related [76] by

$$\gamma_1 \simeq \gamma^* \exp\left(-\frac{C_0}{a^*}\right), \tag{84}$$

$$\gamma_2 \simeq \gamma^* \exp\left(-\frac{2C_0}{a^*}\right), \tag{85}$$

where $\gamma^* \simeq \gamma_0$ and $a^* \simeq 1.5$ Å. Thus the fractional increase in γ_2 with pressure is twice that in γ_1.

D. Minority-Carrier Oscillations

The minority period observed by Soule [90] can be identified with extremal areas in the zone corners [46]. Two situations arise for negative γ_2.

If $\Delta > \epsilon_F$, then the majority-hole surface touches the hexagonal face of the Brillouin zone and a minority-hole pocket is formed, extending along the zone edge until $\epsilon_2^0 < \epsilon_F$. On the other hand, if $\Delta < \epsilon_F$, then the majority-hole surface terminates where $\epsilon_1^0 = \epsilon_3^0$, and no minority-carrier pockets are formed in the corner regions (see Figs. 7 and 8).

TABLE 4

Pressure Derivatives of the Band Parameters of Graphite[a]

	Method of determination			
Derivative	Galvano-magnetic effects between 300 and $420^\circ K$ [76]	Shubnikov-De Haas oscillations $(4.2^\circ K)$ [91]	De Haas-Van Alphen oscillations $(4.2^\circ K)$ [92]	Galvano-magnetic effects $(4.2^\circ K)$ [100]
$(d \log \gamma_1 \gamma_2)/dp$		0.03^5	0.03^6	0.03^7
$(d \log \gamma_1)/dp^b$	0.01^9	$^b 0.01^2$	$^b 0.01^2$	$^b 0.01^2$
$(d \log \gamma_2)/dp^b$	$^b 0.03^8$	$^b 0.02^4$	$^b 0.02^4$	$^b 0.02^4$
$(d \log \epsilon_F)/dp$	$^b 0.03^8$	$^b 0.02^4$	$^b 0.02^4$	$^b 0.02^4$
$(d \log \Delta)/dp$			0.09	

[a] All values calculated assuming $(d \log \gamma_0)/dp = 0$.

[b] Value calculated using the assumption of Arkhipov et al. [76] that $(d \log \gamma_2)/dp = 2 \, [(d \log \gamma_1)/dp]$.

When spin-orbit splitting is taken into account, the areas in the planes $k_z = \pm \pi/C_0$ are extremal (Fig. 8b). For $\Delta > \epsilon_F$ the areas of majority and minority carriers in this plane are given approximately by Eq. (30), allowing Δ to be calculated once ϵ_F is known (Fig. 27a). The spin-orbit coupling λ changes this area by an amount

$$\frac{\delta S}{S_0} \simeq \pm \frac{\lambda \Delta}{\epsilon_F (\epsilon_F - \Delta)} ,$$

(86)

which may be appreciable (~20%) if $\lambda \simeq 10^{-3}$ eV.

The minority-carrier oscillations have since been observed by a number of workers (Table 3). Earlier differences in the minority periods of natural and pyrolytic graphites observed by Williamson et al. [5, 94] have not been substantiated by more recent work (Fig. 28). In a detailed study of the minority periods using oscillations observed in the Hall effect and thermo-electric power, Woollam [93] obtained periods [$\Delta(1/H) \simeq 1.2 \times 10^{-4}$ gauss^{-1}] very close to those obtained for natural crystals by Soule [90] (~1.35 $\times$ 10^{-4} gauss^{-1}). It is possible that the larger periods found by Williamson et al. [5, 94] in pyrolytic graphite [$\Delta(1/H) \simeq 2.24 \times 10^{-4}$ gauss^{-1}] and tentatively by Soule [90] (~2.5 $\times$ 10^{-4} gauss^{-1}) correspond to the minority-hole pocket, whereas the smaller period (larger cross-sectional area) corresponds to the majority-hole extremal area in the planes $k_z = \pm \pi/C_0$ [93] [Fig. 8(b)].

This conclusion is substantiated by measurements of the anisotropy of the periods obtained by tipping the field away from the c-axis direction. Soule [90] showed that the surfaces were closed out to $\theta \simeq 90^\circ$ (Fig. 28) with anisotropy $k_F/\kappa_F \simeq 9$. The ratio obtained by Williamson et al. [5, 94] may be subject to a magnet-calibration error. Woollam [93] finds that the apparent anisotropy ratio increased with increase in the angle θ for measurements up to 87°, amounting to 13.5 at this angle. The behavior is easily understood if this carrier is associated with the majority-hole surface at H (see Fig. 8b).

The anisotropy ratio for the minority surface varies with the parameter Δ. The observed periods correspond to a much larger value for Δ than that given by the magnetoreflection technique, which correspond to anistropy values for the minority-hole surface less than about 4 or 5 (Fig. 27b).

If the assignment of the carriers is correct, the minority period should decrease with acceptor doping, corresponding to an increase in the extremal

(a)

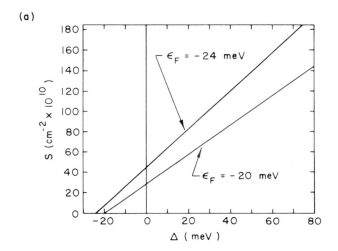

(b)

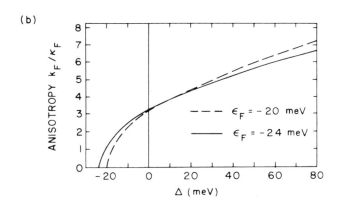

FIG. 27. (a) The variation in the cross-sectional area of the Fermi surface in the planes $\phi = \pm \pi/2$. When spin-orbit interaction is taken into account, two extremal areas result. Since the spin-orbit-interaction parameter $\lambda_{3,3}$ is less than 1 meV, the areas should not be modified significantly
from the average value shown. (b) The ratio of the principal wavevectors
for the Fermi surface of the minority-carrier pocket, plotted as a function
of the overlap parameter.

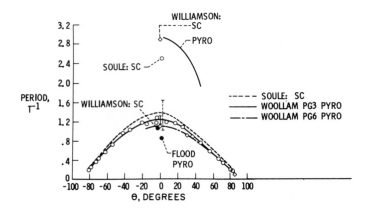

FIG. 28. Periods of oscillation observed for the minority carriers in graphite by several workers. The variation in period with θ, the angle between the c-axis direction and the magnetic field, is also shown. Figure from Woollam [93]. Data from Flood [101], Soule [90], Williamson et al. [5], and Woollam [93].

cross-sectional area as the Fermi energy is lowered. A large decrease in the period has been observed by Soule [90] for a doping level of only 13 ppm, which suggests that the depression of the Fermi level is not the only mechanism responsible for the change if the carriers are located near H (Fig. 26).

Anderson et al. [92] also measured the pressure dependence of the minority-carrier period, obtaining a relatively large pressure coefficient for the parameter Δ (d log Δ/d log P $\approx$ 0.09 kilobar^{-1}). It is unfortunate that the only independent check of the pressure derivative of the number of minority carriers observed in the Hall coefficient [102] suggests that the carriers originate from different parts of the Brillouin zone (see Section VII.H).

It is not certain at the moment that the long periods in the de Haas van-Alphen Effect can be attributed to carriers near point H. It is possible that they could be located in the region of trigonal warping between the majority hole and electron surfaces. The carriers in this region have not received the attention they deserve.

E. Non-oscillatory Susceptibility

The susceptibility of graphite parallel to the principal axis is large and negative (-21.5×10^{-6} emu/g) at room temperature, increasing with decreasing temperature [103]. In the simplest model diamagnetism arises if the energy of the electron assembly is increased when the electrons occupy the energy levels produced by the magnetic field. By using a two-dimensional model of the energy bands, McClure [70] was able to show that the correct order of magnitude for the susceptibility could be obtained once the degeneracies of the bands were taken into account.

In a later paper, using the three-dimensional band structure, McClure [66, 104] obtained the magnetic energy levels already discussed in Section IV. B and fitted the experimental curves of the susceptibility. The low-temperature susceptibility is approximately equal to $-3\gamma_0^2 \times 10^{-6}$ emu/g, affording a sensitive estimate of the parameter γ_0. McClure obtained a γ_0 value of 2.8 ± 0.1 eV, which is in reasonable agreement with other estimates. The dependence of the susceptibility on temperature can then be used to obtain a value for $\gamma_1 = 0.27$ eV, which is lower than other estimates (Table 6). It has been pointed out, however, that the adoption of a negative value for γ_2 affects the details of the susceptibility calculation and might lead to closer agreement [20].

In the above calculation the small Pauli paramagnetic susceptibility was ignored, although this has been calculated [98]. Similarly the contribution from the ion cores is very small ($\chi = -0.33 \times 10^{-6}$ emu/g) [98].

At boron doping levels of approximately 140 ppm the electrochemical potential shifts to the conduction-band edge, producing a knee in the susceptibility-versus-doping-concentration diagram [98]. The curve can be fitted well if the ionization efficiency of the boron atoms is assumed to be 67%. It is not known whether the sign of γ_2 affects this result, since few details of the calculation have been given [98].

F. High-Field Oscillatory Effects

The interpretation of high-field oscillatory effects is important since the experiments afford an independent method of determining the sign of γ_2. The high-field limit of graphite can be readily defined in terms of the field required to place the m = 1 level for the majority carriers into coincidence with the Fermi energy that is, ~70 kG for electrons and ~35 kG for holes. The coincidences may be located readily (but not identified as arising from holes or electrons), using the condition that the magnetoconductivity-tensor component σ_{xx} should give a sharp maximum. In graphite, since $\sigma_{xx} \gg \sigma_{xy}$ (see Section VII), this implies that the measured resistivity ρ_{xx} is at a minimum (Fig. 29). The minima may be related to particular pieces of the Fermi surface, as discussed in Section VI.C.

The sign of the charge carriers can be determined from the behavior of the coincidences of the Hall effect and magnetothermopower. Woollam [21, 96] used a theory due to Argyres [106], indicating that the Hall resistivity ρ_{yx} is at a minimum value for hole coincidences and a maximum value for electron coincidences. This enables the smaller period oscillation associated with the larger cross section at point K to be identified as electron-like (Fig. 29), supporting the assignment of Schroeder et al. [20]. Further confirmation of this assignment comes from the observation of sharp maxima in the thermoelectric power $S_{xx}(H)$ (Fig. 30) for holes and minima for electrons in agreement with theory [21, 96].

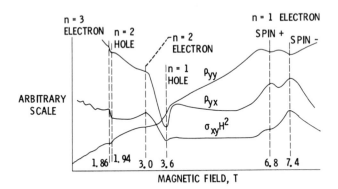

FIG. 29. Variation of the magnetoresistance and Hall-tensor components ρ_{yy} and ρ_{yx} in the high-field region for graphite. The off-diagonal conductivity-tensor component σ_{xy} is also shown. Minima in ρ_{yx} and σ_{xy} correspond to hole carriers, and maxima to electron carriers. From Ref. [96].

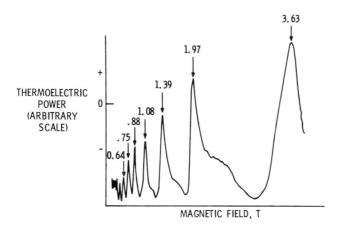

FIG. 30. Sharp maxima in the thermoelectric power for graphite, indicating coincidences of the Landau levels for holes with the Fermi energy.
From Ref. [96]. Numbers refer to field values in tesla.

It is interesting to note that the oscillations obtained by Berlincourt and
Steele [86] from the Hall coefficient showed sharp minima for the hole oscillations in agreement with Woollam's findings, but his results were not
analyzed in detail.

G. Spin-Orbit Splitting in Magnetic Fields and the g Factor

Wagoner [107] showed that the g factor in graphite has a remarkable
anisotropy. The parameter was obtained from measurements of the con-
duction-electron spin resonance in natural crystals, where the resonance
condition can be written as follows:

$$\hbar\omega = g\mu_B H, \tag{87}$$

where μ_B is the Bohr magneton and ω is the frequency of the incident radiation. The g values corresponding to magnetic field parallel and perpendicular
to the c-axis (g_3 and g_1) were 2.0495 and 2.0026, respectively, at room
temperature, and 2.127 and 2.0026 at 77°K. This compares with the free-
electron value of 2.0023.

Wagoner [107] concluded that the large shift in g_3 is a result of the perturbing effects of nearly degenerate states on a given state near the Fermi surface. A theory was worked out by McClure and Yafet [47], who concluded that the g shift vanishes if the interaction between layers is neglected. The data could be fitted if the unusually large value of $\Delta = 0.1$ eV was chosen.

More recently Woollam [79] observed splittings in the $m = 1$ and $m = 2$ electron magnetic energy levels, and $m = 1$ hole level (Fig. 31). Since the De Haas-Van Alphen effect picks out orbits at particular points in the Brillouin zone, the g shifts measured in this way can be compared more directly with theory. Making allowance for the shift of the Fermi energy with magnetic field, the g shift can be calculated from the data (Table 5). The theory of McClure and Yafet [47] obtains a value for g/λ, where λ is the spin-orbit-interaction parameter. Using their value of $\lambda = 3 \times 10^{-4}$ eV (see Section III. G), agreement between calculated and experimental g values is not satisfactory. This may be due to the neglect of the parameter γ_3 in the theory.

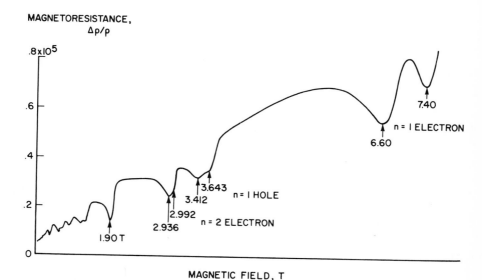

FIG. 31. Variation of the resistivity tensor component ρ_{xx} with magnetic field, showing the splitting of the Landau levels due to spin-orbit interaction; PG5 graphite at 1.1°K. From Ref. [79]. Numbers refer to field values in tesla.

TABLE 5

Comparison of g Shifts[a] Obtained from Theory and Experiment[b]

Landau level	Theory $\delta g/\lambda$ (x 10^4)	Experiment δg	Experiment $(\delta g/\lambda$ (x 10^4)
m = 1, electron	-0.05	0.7 ± 0.4	+0.2 ± 0.1
m = 2, electron	-0.07	0.0 ± 0.3	0 ± 0.1
m = 1, hole	+0.06	3.0 ± 0.5	+1.0 ± 0.1

[a]Splitting of high-field Landau levels [79].
[b]Data from Ref. [48].

H. Effects Related to Plasmons

Various modes of plasma oscillation may be set up in crystals, depending on the band structure. In compensated semimetals, Alfvén waves [109] may propagate, in which hole and electron movements are $180°$ out of phase with each other, but of equal amplitude. Alfvén-wave propagation occurs in compensated conductors when $\omega_p \gg \omega_c \gg \omega$ or r^{-1}, where ω_p, ω_c, ω, and r^{-1} are the plasma, cyclotron, signal, and collision frequencies, respectively. It follows that the effect can be observed only at very low temperatures. The response of the crystal to the applied field can be discussed in terms of the effective dielectric constant

$$K = K_0 + \frac{4\pi c^2}{H^2} \sum_{i=1}^{N} \int_{\substack{\text{over Fermi} \\ \text{surface}}} n_i(k_z) \, m_i^*(k_z) \, dk_z, \tag{88}$$

$$K = K_0 + \frac{4\pi c^2}{H^2} D(n, m^*) \tag{89}$$

where $D(n, m^*)$ is the carrier mass-density function and is equal to $(n_e m_e^* + n_h m_h^*)$ for a compensated semimetal with spheroidal bands. Surma,

Furdyna, and Praddaude [110] have observed Alfven-wave propagation in synthetic and natural crystals with the field parallel (Fig. 32) and at an angle (Fig. 33) to the c-axis. With the sample in the center of the cavity, the boundary conditions at the sample faces allow the mass-density function to be expressed in terms of the period $\Delta(1/H)$ of the interference pattern. If

$$\ell\lambda = d, \tag{90}$$

where ℓ is an integer, λ is the wavelength of the beam, and d is the thickness of the specimen, then

$$D(n, m^*) = 10^{-10} \, \pi \, [\omega d \, \Delta \, (1/H)]^{-2} \; (g/cm^3). \tag{91}$$

The results obtained $[D_{exp} = (3.0 \pm 0.3) \times 10^{-10} \; g/cm^3]$ for H parallel to the c-axis agreed well with the value calculated by using the Slonczewski-Weiss [12] band model ($\gamma_3 = 0$): $D_{theor} = 3.25 \times 10^{-10} \; g/cm^3$. Results

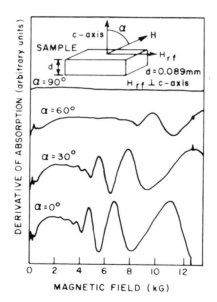

FIG. 32. The derivative of absorption as a function of magnetic field for two specimens of pyrolytic graphite of different thickness. The large oscillations represent the Alfvén wave-interference pattern; cyclotron resonance appears in the low-field region. The sharp spike at 12.5 kG is an EPR marker. From Ref. [110].

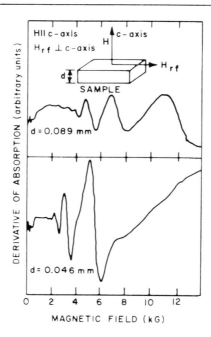

FIG. 33. Anisotropy of the Alfven-wave propagation in pyrolytic graphite. The pattern is essentially determined by the projection of B on the c-axis. From Ref. [110].

obtained with the field tipped away from the c-axis (Fig. 33) suggested that the interference pattern depended only on the normal component of field H cos a, so that the plasma behaved as though it were two-dimensional in nature.

Saunders [19] has pointed out the potential importance of plasma effects in graphite. Recent calculations on the growth of two-stream instabilities in electron-hole plasmas in graphite [111] may well lead to the exploitation of plasma effects in devices. The instability arises when radio-frequency fields stimulating plasma oscillations are of sufficient amplitude to overcome intrinsic damping effects. These authors concluded that the most favorable condition for the instability occurs at a frequency of 3×10^{13} Hz ($\lambda = 10\mu$) for an electric field of 160 V/cm.

Low-energy plasma resonances can be observed in reflection and absorption studies. Earlier measurements in the infrared region (1.5-32 μ) [112] attributed a minimum in the emissivity at about $4\,\mu$ to a transition $\epsilon_2^0 \rightarrow \epsilon_3^0$ at point K, obtaining a value $2\gamma_1 = 0.28$ eV, which is much smaller than other estimates. McClure [17] proposed that the minimum comes when the frequency of the incident radiation exceeds that of the plasma frequency. At the temperature of the measurement (523°K) the plasmon energy [Eqs. (38) and (39)] is increased over its value at 0°K because of the increase in the number of free carriers ($\hbar\omega_p \simeq 0.3$ eV). Absorption measurements [113] confirm that $\epsilon_2^0 \rightarrow \epsilon_3^0$ transitions at point K occur for photon energies of about 0.8 eV, giving a value of $\gamma_1 = 0.4$ eV, in agreement with other estimates. Further measurements at very low photon energies (0.13-0.006 eV) [114] tentatively assigned weak peaks in the reflectance at 0.14 and 0.012 eV to the compressional plasma modes discussed in Section III. J. This gives an effective-mass anisotropy $m_{33}^*/m_{11}^* = 139$, in close agreement with other estimates.

Other plasma resonances have been reported for higher energy values. Electron-energy-loss experiments using electron beams parallel to the c-axis indicate plasma energies of about 6.3 and 25 eV [115] or 4.6 and 19 eV [116, 117]. The lower energy loss ($\Delta\epsilon_\pi$) can be identified with the π-electron system [118], affording an estimate of the parameters γ_0 and γ_1, since

$$\Delta\epsilon_\pi = \hbar\left(\frac{4\pi n e^2}{m^*}\right)^{1/2}\left\{1 - \frac{m^* a_0^2 \gamma_0}{3\hbar^2}\left[1 + \frac{\gamma_1}{4\gamma_0}\left(\frac{c_0}{a_0}\right)^2\right]\right\}^{1/2}, \tag{92}$$

$\gamma_0 = 1.63$eV, $\gamma_1 = 0.42$eV [118]. The higher energy loss $\Delta\epsilon_{\sigma\pi}$ may be related to the σ and π electrons (four electrons per atom). These plasmons can also be identified from other optical studies [119-122]. Zeppenfeld [116, 117] has also studied the energy loss for an electron beam incident at an angle with the c-axis. Greenaway and co-workers [120] then reinterpreted the low-energy loss $\Delta\epsilon_{\sigma\pi} = 6.8$ eV [116, 117] for an incident-beam perpendicular to the c-axis as resulting from interband transitions, so that the energy shift does not result from plasmon anisotropy, as suggested by Zeppenfeld [116, 117]. The identification of interband transitions has been made possible by investigations of the band structure for electron energies far from the Fermi energy [39, 62, 63] (see Section III. K).

I. Summary of Experimental Values of Band Parameters for Graphite

Table 6 lists the band parameters and experiments that have given information about their magnitudes. At the moment there is no one set of band parameters that gives agreement with all experiments. It is also difficult to assess values quoted by various workers since each adopts a separate approximation. For instance, the De Haas-Van Alphen periods are related to the band parameters γ_0, γ_1, γ_2, γ_4, γ_5, as can be seen from Eqs. (17) and (18). The experiment is usually quoted as giving results for γ_2, but values are computed from the data by using different values for γ_4, γ_5. Furthermore the maximal cross-sectional area of the electrons is dependent also on γ_3 [not included in Eqs. (17) and (18)] (Table 8).

The largest uncertainty at present comes from γ_3, which may be estimated from the line shape found in the magnetoreflection experiment. Most recent results [45] indicate that its value is as high as 0.29 ± 0.02 eV. For values greater than 0.2 eV the effect of γ_3 may no longer be treated as a perturbation, and the calculation of the energy surfaces and dynamical properties of the carriers becomes exceedingly complex. Such calculations are presently being made but are not completed [48]. Values,quoted in Table 6 are tentative. It is instructive to compare tables of quoted values in past review articles. It is clear that error limits have generally been evaluated with more hope than realism.

The Fermi energies quoted in Table 6 were not directly measured but rather calculated by using the Slonczewski-Weiss [12] band model. Values for the effective mass of the electrons at extremal points are given in Table 7.

VII. TRANSPORT PROPERTIES IN THE BASAL PLANE

A. Phenomenological Equations for the Transport Coefficients

If measurements are made on a block of material with primary electric-current density j and heat-current density q, then an electromotive force E and temperature gradient ∇T are set up, related in the limit of small current flow by the linear equations

TABLE 6

Experimental Values of the Band Parameters in Graphite

Para-meter	Experiment	Value (eV)	Refs.
γ_0	Nonoscillatory susceptibility	2.8	66
	Energy loss from electron beam	1.63	118
		2.17	116
	Magnetoreflection, high field, point H	3.21 ± 0.05	44
γ_1	Nonoscillatory susceptibility	0.27	66
	Magnetoreflection, high quantum levels, point K	0.400 ± 0.005	44
	Energy loss from electron beam	0.42	118
		0.47	116
	Infrared reflectivity	~0.4	121
	Photoemission and secondary-electron emission	0.42	122
γ_2	DHVA periods	-0.015 to -0.020	5,22, 84-88, 90-93
γ_3	Magnetoreflection, low quantum levels, point K	0.29 ± 0.02	45
	De Haas-Van Alphen period, majority electron	≤ 0.15	84
γ_4	Magnetoreflection, low quantum levels, point K	0.20	65
		0.25 ± 0.02	20
	g Shift, electron spin resonance	0.28	47
	Ratio of De Haas-Van Alphen periods	Dependent on γ_1, γ_2, γ_3, ~0.1 to 0.2	5,22, 84-88, 90-93
γ_5	Not measured; $\gamma_5 = \gamma_2$	~0.02	123
Δ	De Haas-Van Alphen period, majority hole	0.14	84
	De Haas-Van Alphen periods, minority carriers	0.005 ± 0.001	5,94
	Magnetoreflection, low field, point H	0.009 ± 0.003	93
	Temperature dependence of the g factor	~0.1	47
ϵ_F	Number of holes = number of electrons	-0.020 to -0.024	44

TABLE 7

Effective Masses of Graphite Deduced from Experimental Measurements

Method of determination	Effective mass m*						
	Majority-electron, basal, $(m_{11}^*)_{el}$	Majority-hole, basal, $(m_{11}^*)_h$	Minority-carrier, basal, $(m_{11}^*)_{min}$	Majority-electron, c-axis $(m_{33}^*)_{el}$	Majority-hole, c-axis $(m_{33}^*)_h$	Minority-carrier, c-axis, $(m_{33}^*)_{min}$	Ref.
Field and temperature dependence of magneto-resistivity and magneto-conductivity	$(0.057 \pm 0.002)m_0$	$(0.039 \pm 0.001)m_0$	--	$14m_0$	$5.7m_0$		84^a
Field dependence of de Haas-Van Alphen oscillations	--	--	$0.0023m_0$	--	--	$0.017m_0$	90
	$(0.057 \pm 0.002)m_0$	$(0.039 \pm 0.001)m_0$	$(0.004 \pm 0.0004)m_0$	--	--	--	5
Field dependence of resonance absorption in cyclotron resonance	$(0.061 \pm 0.002)m_0$	--	--	--	--	--	82
Magneto-reflection point K.	$(0.061 \pm 0.002)m_0$	--	--	--	--	--	20

[a] See also Ref. [88].

TABLE 8

Effect of γ_3 on De Haas-Van Alphen Periods and Carrier
Effective Masses[a]

γ_3(eV)	P hole $(10^{-5}$ gauss$^{-1})$	m* hole, m_0	P electron $(10^{-5}$ gauss$^{-1})$	m* electron, m_0
0.00	2.08	0.0397	1.57	0.0527
0.10	2.08	0.0397	1.53	0.0529
0.15	2.07	0.0397	1.56	0.0533
0.20	2.07	0.0397	1.62	0.0544
0.25	2.06	0.0397	1.72	0.0567
0.30	2.06	0.0397	1.92	0.0627

[a]Data adapted from Ref. [5]. Values calculated for $\gamma_0 = 3.13$ eV, $\gamma_1 = 0.40$ eV, $\gamma_2 = \gamma_5 = -0.0185$ eV, $\gamma_4 = 0.25$ eV; $\Delta = 0.005$ eV, $\epsilon_F = -0.022$ (assumed independent of γ_3).

$$j_i = \sigma_{ij}(H)\underline{E}_j + M_{ij}(H)\ \underline{\nabla}_j T \quad (i, j = 1, 2, 3), \tag{93}$$

$$q_i = N_{ij}(H)\underline{E}_j + L_{ij}(H)\ \underline{\nabla}_j T \quad (i, j = 1, 2, 3). \tag{94}$$

In this treatment the measured emf or electrochemical field is taken to be the sum of the electrostatic field $\underline{E}^*$ and the gradient of the electrochemical potential:

$$\underline{E} = \underline{E}^* - \frac{1}{e}\ \underline{\nabla}\eta, \tag{95}$$

where $e = -|e|$ for electrons and $e = +|e|$ for holes. It is preferable to work theoretically with the conductivity coefficients σ, M, N, L since the contributions from different groups of carriers are directly additive. However, in an experiment the currents $[J_i]$ are applied and temperature and potential differences are measured. The measured coefficients of interest are included in Table 9. We are concerned only with isothermal coefficients.

TABLE 9

Definition of the Galvanomagnetic and Thermomagnetic Coefficients
Discussed in the Text

Name of effect	Symbol	Definition
Resistivity[a]	ρ_{xx} or ρ_{11}	$\rho_{xx} = \dfrac{E_x}{j_x} = \dfrac{\Delta V_x s_y s_z}{J_x s_x}$
Magnetoresistance coefficient[a]	M	$\dfrac{\Delta \rho}{\rho_0 H^2}$
Hall resistivity[a]	ρ_{yx} or ρ_{21}	$\rho_{yx} = \dfrac{E_y}{j_x} = \dfrac{\Delta V_y s_z}{J_x}$
Hall coefficient[a]	R or R_H	$R = \dfrac{\rho_{yx}}{H_z}$
Thermoelectric power, or Seeback coefficient[b]	S_{xx} or S_{11}	$S_{xx} = \dfrac{E_x}{\partial T/\partial_x} = \dfrac{\Delta V_x}{\Delta T_x}$
Nernst resistivity[b]	S_{yx}	$S_{yx} = E_y \div \dfrac{\partial T}{\partial x}$
Nernst coefficient[b]	Q	$Q = \dfrac{S_{yx}}{H_z}$

[a]Auxiliary conditions for isothermal effect:

$$J_y = \frac{\partial T}{\partial x} = \frac{\partial T}{\partial y} = 0.$$

[b]Auxiliary conditions for isothermal effect:

$$J_x = J_y = \frac{\partial T}{\partial y} = 0.$$

(s_x, s_y, s_z represent the sample dimensions.)

In any solid it is preferable to perform experiments in which the current is applied along principal directions. If the current flow is parallel to atomic planes, then the symmetry of the crystal ensures the following relationships:

$$\sigma_{xx}(\underline{H}) = \sigma_{xx}(-\underline{H}) = \sigma_{yy}(\underline{H}), \tag{96}$$

$$\sigma_{xy}(\underline{H}) = -\sigma_{xy}(-\underline{H}) = -\sigma_{yx}(\underline{H}), \tag{97}$$

$$M_{xx}(\underline{H}) = M_{xx}(-\underline{H}) = M_{yy}(\underline{H}), \tag{98}$$

$$M_{xy}(\underline{H}) = -M_{xy}(-\underline{H}) = -M_{yx}(\underline{H}) \tag{99}$$

By using these relationships, the Hall coefficient ρ_{yx}/H_z, measured resistivity ρ_{xx}, thermoelectric power S_{xx}, and Nernst coefficient S_{yx}/H_z can be obtained in terms of σ and M or the conductivity coefficients obtained from the measured coefficients (Tables 10 and 11). The relationships between ρ and σ were first used by Soule [87] for graphite, and those between S and σ, M, by Spain and Platz [24].

TABLE 10

Relationships between Measured Coefficients and
Conductivity Tensor Components

Coefficient	Symbol	Expression
Resistivity	$\rho_{xx} = \rho_{yy} \equiv \rho$	$\dfrac{\sigma_{xx}}{\sigma_{xx}^2 + \sigma_{xy}^2}$
Hall coefficient	$R = \dfrac{\rho_{yx}}{H} = -\dfrac{\rho_{xy}}{H}$	$\dfrac{1}{H_z}\left(\dfrac{\sigma_{xy}}{\sigma_{xx}^2 + \sigma_{xy}^2}\right)$
Thermoelectric power	$S_{xx} = S_{yy}$	$-\left(\dfrac{M_{xx}\sigma_{xx} + M_{xy}\sigma_{xy}}{\sigma_{xx}^2 + \sigma_{xy}^2}\right)$
Nernst coefficient	$Q = \dfrac{S_{yx}}{H_z} = -\dfrac{S_{xy}}{H_z}$	$-\dfrac{1}{H_z}\left(\dfrac{M_{xx}\sigma_{xy} - M_{xy}\sigma_{xx}}{\sigma_{xx}^2 + \sigma_{xy}^2}\right)$

TABLE 11

Relationship between Conductivity Tensor Components
and Measured Coefficients[a]

Tensor component	Relationship	Alternative relationship
σ_{xx}	$\dfrac{\sigma}{1 + (R\sigma H)^2}$	--
σ_{xy}	$\dfrac{\sigma(R\sigma H)}{1 + (R\sigma H)^2}$	$R\sigma H \sigma_{xx}$
M_{xx}	$\dfrac{\sigma}{1 + (R\sigma H)^2}(RQ\sigma H^2 - S)$	$QH\sigma_{xy} - S\sigma_{xx}$
M_{xy}	$-\dfrac{H\sigma}{1 + (R\sigma H)^2}(Q + SR\sigma H)$	$-(QH\sigma_{xx} + S\sigma_{xy})$

[a]Here the measured conductivity σ is reciprocally related to the measured resistivity ($\sigma = 1/\rho$), $S_{xx} \equiv S$, $S_{yx} \equiv Q$.

B. Calculation of the Transport Coefficients by Using the Boltzmann Transport Equation

Under the action of electric fields and thermal gradients the electrons gain energy from the fields. Eventually a steady state is reached, when the energy losses due to collisions balance the energy gained from the fields. It is convenient to describe the electrons by a distribution function $f(\underline{k}, \underline{r})$ expressing the normalized probability of finding an electron with wavevector $\underline{k}$ at the position coordinate $\underline{r}$. The electric and thermal current densities can then be written as follows:

$$\underline{j} = \frac{e}{4\pi^3}\int \underline{v}\, f\, d^3k, \tag{100}$$

$$\underline{q} = \frac{1}{4\pi^3}\int \underline{v}\,(\epsilon - \eta)f\, d^3k \tag{101}$$

If the energy-wavevector relationship is parabolic,

$$\epsilon = \frac{\hbar^2}{2} k_i \bar{\bar{M}}_{ij} k_j, \tag{102}$$

where $\bar{\bar{M}}_{ij}$ is the reciprocal effective-mass tensor

$$\bar{\bar{M}}_{ij} = \frac{1}{\hbar^2} \frac{\partial^2 \epsilon}{\partial k_i \partial k_j} \tag{103}$$

then the distribution function in the steady state can be found from the Boltzmann transport equation. If

$$f = f_0 - \phi(\underline{k}) \frac{\partial f_0}{\partial \epsilon}, \tag{104}$$

where f_0 is the equilibrium distribution function (see, for example, Ref. [124]),

$$\phi(k) = \frac{-e\tau(\underline{k})}{\hbar} \text{grad}_k \epsilon \frac{\underline{F} + (e\tau/c)(\bar{\bar{M}})\underline{F} \wedge \underline{H} + (e\tau/c)^2 [\underline{F} \cdot \underline{H} (\bar{\bar{M}})\underline{H}/|\bar{\bar{M}}|]}{1 + (e\tau/c)^2 [(\bar{\bar{M}})\underline{H} \cdot \underline{H}/|\bar{\bar{M}}|]} \tag{105}$$

where $\bar{\bar{M}}$ is the effective mass tensor, $|\bar{\bar{M}}|$ is its determinant, τ is the relaxation time for scattering, and

$$\underline{F} = \underline{E} + \frac{1}{eT} (\epsilon - \eta) \underline{\nabla} T, \tag{106}$$

so that thermal and electrical effects are separable. The last term in the numerator of Eq. (105) relates to effects that are not normally measured ($\underline{H}$ not perpendicular to $\underline{F}$) and can be neglected.

This equation, which may be generalized to the case of anisotropic relaxation times [124], is valid provided that the following criteria are satisfied:

1. The time between collisions, τ, must be long compared with the duration of the collision process, Δt; that is,

$$\Delta t \simeq \frac{\hbar}{kT} \ll \tau. \tag{107}$$

This condition can alternatively be written as

$$\lambda \ll \ell, \tag{108}$$

where λ is the wavelength of electron and ℓ is the mean free path. It is satisfied for collision processes that control the _basal_ conductivity at all temperatures of interest.

2. If there is a temperature gradient in the material, the electrons should be in thermal equilibrium with the lattice in any small region. This condition breaks down only at very low temperatures in very perfect crystals where the mean free path is exceptionally long.

3. In magnetic fields the formula is valid well above the normal condition

$$\frac{eH}{m^*c} \tau \equiv \omega_c \tau \lesssim 1. \tag{109}$$

The condition for applicability is that quantum effects should not dominate, that is,

$$\hbar\omega_c < kT \quad \text{or} \quad \eta. \tag{110}$$

This restricts the magnetic field region to below 30 kG for temperatures above 77°K, and correspondingly lower fields as the temperature is lowered.

C. Relaxation Time for Scattering

The formation of the Boltzmann transport equation discussed in the preceding section assumed that the collisions of the carriers could be adequately described by a relaxation time, depending in general on the wavevector of the carrier. This implies that if all external fields are switched off, the distribution function f returns to its equilibrium value f_0 according to

$$\frac{\partial f}{\partial t} = \frac{f - f_0}{\tau(\underline{k})}. \tag{111}$$

Theoretical treatments of the conductivity in graphite have in general assumed that the predominant scattering mechanism above about 50°K comes from collisions with lattice vibrations (phonon–electron scattering). Since this interaction becomes very weak at low temperatures, scattering from defects such as grain boundaries, ionized impurities, vacancies, and interstitials predominates. The total relaxation time is usually expressed by using Mathiesson's rule [124].

$$\frac{1}{\tau_{\text{total}}} = \frac{1}{\tau_{\text{el-ph}}} + \frac{1}{\tau_{\text{def}}}, \tag{112}$$

where the relaxation time from scattering by defects (τ_{def}) may itself be a similar sum of terms from several mechanisms.

The relaxation time from scattering by the lattice vibrations $\tau_{\text{el-ph}}$ can be calculated from first principles. The essential idea is that a lattice wave produces a local change in potential (the deformation potential -D) which scatters the electron. The probability of scattering, which is reciprocally related to the relaxation time, may then be calculated by summing all possible collisions over the statistical distribution of lattice vibrations and electrons. Only collisions are allowed for which the quasi-momentum- and energy-conservation laws are satisfied. For a scattering event in which an electron of wavevector $\underline{k}_1$ is scattered into a state $\underline{k}_2$ by a phonon of wavevector $\underline{q}$ and energy $\hbar\omega(\underline{q})$,

$$\underline{k}_1 \pm \underline{q} = \underline{k}_2 + \underline{G}, \tag{113}$$

$$\epsilon_1 \pm \hbar\omega(\underline{q}) = \epsilon_2. \tag{114}$$

The + sign refers to annihilation, the - sign to creation of a quantum of lattice vibration (phonon). Collisions for which the reciprocal lattice vector $\underline{G}$ is zero are termed Normal collisions, those for which $\underline{G} \neq 0$ are termed Umklapp collisions. Umklapp collisions may be conveniently neglected for graphite if the calculation is made in the extended-zone scheme. This ensures that for all temperatures of interest, equivalent collisions can be treated as Normal processes.

It can be shown that the probability of scattering is proportional to the density of states into which scattering occurs and to the number of phonons excited at the temperature (T) [125]. Assuming that intravalley-scattering processes predominate, and since the energy density of states for small electron and hole energies is lower for holes than for electrons, this implies that the average relaxation time for scattering holes should be larger than that for the electrons. However, intervalley scattering is allowed from collisions with c-axis propagating modes even for low temperatures (T $\leq 50^{\circ}$K) since the characteristic temperature is less than 100°K. (For a review of lattice properties of graphite see Ref. [126]). Since the basal area of the Fermi surface is small, the scattering by basal propagating modes involves very

small phonon wavevectors. For temperatures above the degeneracy temperature ($kT \geq \epsilon_F$) such modes may also produce intervalley scattering [23].

For temperatures above about $50^\circ K$ the number of phonons of both
propagation directions depends directly on the temperature (classical
statistics are valid) and the relaxation time may sensibly be written

$$\tau = \frac{\tau_0}{T} \, \epsilon^p \, w(\phi) \qquad (\phi = \frac{k_z C_0}{2}) \tag{115}$$

For a typical semiconductor the density of states is proportional to
$\epsilon^{1/2}$, therefore $p = -1/2$. For graphite, $p = 0$ in the extreme of the cylindrical band model and takes an intermediate value for the Slonczewski-Weiss
model. Arkhipov et al. [76] have used a relaxation time of this form, with
$w(\phi)$ taking the simple form $w(\phi) = \cos\phi$, with $p = 0$, the exponent in T being allowed to vary (see also Ref. [127]). Kechin et al. [25, 100] have used the
more general form $w(\phi) = \cos^S (\phi)$, where the parameter S was fitted to
experiment.

A more detailed calculation has been made by Sugihara and Sato [128],
who divided the lattice vibrations of interest into longitudinal modes propagating along the basal planes (mode 1) and perpendicular to them (mode 2). In
the limit of low phonon wavevector for mode 1 these modes are almost independent [129, 130], but they have very different characteristic tempera-
tures ($\theta_1 \simeq 2000$, $\theta_2 \simeq 100^\circ K$) and velocities ($s_1 \simeq 2 \times 10^6$, $s_2 \simeq 4 \times 10^5$ cm/
sec). Good agreement with the measured conductivity was obtained if the
deformation-potential constants for the modes were taken as $D_1 = 21$, D_2
$= 3.5$ eV (Table 12) (see also Ono and Sugihara [77]).

The effect of the transverse modes was not included. This is usually a
good approximation for a metal, in which, since to a first-order approximation the shear modes produce no change in volume, the deformation potential
is very small. However, this is no longer true even in a metal when the
Fermi surface touches the zone boundary [131]. In graphite one might ex-
pect the deformation potential for the shear and compressional waves to be
comparable as a result of the directional nature of the bonds.

The deformation potential $D_1 \simeq 21$ eV may be compared with a similar
value found in diamond and other covalent semiconductors [51]. Since the
conduction electrons in graphite are associated with much weaker van der
Waals forces, this value seems high. An ab initio calculation is required.

TABLE 12

Deformation-Potential Constants for Electron-Phonon
Scattering in Graphite[a]

D_1 (eV)	D_2 (eV)	Ref.
21	3.5[b]	128
16.2	3.7	77
27.9[c]	0[c]	42
30[d]	0	105

[a]Longitudinal modes only considered.

[b]Values corrected by a factor of $\sqrt{2}$ to agree with definition of D used in Refs. [42] and [77] (see note in Ref. [77]).

[c]Effect of trigonal warping included.

[d]Corrected value (see Ref. [17]).

Further calculations of the deformation-potential constants are included in Table 12. Overall agreement is reasonably good. In the later paper of Ono and Sugihara [42] in which the effect of trigonal warping was considered, the result that $D_2 \simeq 0$ is very different from the previous results, since the scattering probability is proportional to $(D_i/s_1)^2$. With the values quoted in Refs. [77] and [128], the ratio of scattering from the two modes is roughly equal. The inclusion of transverse modes in the calculation and allowance for the effect of interelectronic collisions could change the ratios quoted and reduce the deformation potentials to more reasonable values.

In assessing the effect of the trigonal warping of the bands on the relaxation time, Ono and Sugihara [42] concluded that at higher temperature the magnitude of r is reduced by about 20%, whereas the dependence of r on energy is not changed. They also arrived at the surprising result that the average relaxation time of the electrons near point K becomes longer than that for the holes. Since it might be thought that low-angle scattering would become more effective in controlling the conductivity in warped energy surfaces, the reason for this phenomenon is not immediately clear.

One feature of the scattering processes in graphite that has not received attention is the difference between elastic and inelastic scattering by
phonons. Inelastic processes may be quite effective in bringing a perturbed
electron distribution back into a quasi-equilibrium state where the net momentum, and hence current, is zero. However, without inelastic processes,
the temperature of the electron gas would increase above that of the lattice
[132]. The basal propagating modes that interact with the electrons have
little energy, whereas the interacting c-axis propagating modes, with wavevector q_z, are not restrained to small values by the momentum-conservation
law, Eq. (113), and will have an average energy kT. This is not small compared with the electrochemical potential, so that this inelastic process is
very important in maintaining the electron temperature in equilibrium with
the lattice.

The problem of elastic and inelastic scattering has immediate relevance
to the effects of carrier-carrier scattering in graphite. Such effects can be
conveniently separated into three parts: electron-electron (e-e), hole-hole
(h-h), and electron-hole (e-h). The e-e and h-h processes are of importance
only to second order in the scattering, since they may only modify the effects of other scattering processes (e.g., electron-phonon, electron-ionized
impurity scattering). The effect of such interelectronic scattering processes
is to randomize the electron distribution, subject to the condition that no
net momentum loss may occur. Interelectronic collisions are expected to
predominate if the carrier density is greater than [133]

$$n \approx \frac{1}{4\pi} \frac{\eta^{3/2} \, m*^{3/2} \, s^2 K^2}{kTe^4 \tau_{e-ph}} \approx 10^{14} \ cm^{-3} \ at \ 300°K. \tag{116}$$

Therefore these collisions must be important in graphite. Herring [134]
and Keyes [135] have calculated the effects that such processes may have on
the conductivity and galvanomagnetic coefficients, respectively. In the limit
that interelectronic collisions predominate (i.e., τ for interelectronic scattering is much less than the scattering due to all other processes) the effects
may be quite appreciable.

Electron-hole scattering may affect the properties to first order. When
a field is applied to a crystal, electrons and holes diffuse in opposite directions. Electron-hole scattering may therefore transfer net momentum from

one carrier to the other, thus destroying current (mutual-drag effect). The treatment of this effect is complicated [136, 137].

Sugihara [138] has calculated the resulting change in the minority-carrier mobility in graphite. In this calculation the mutual-drag effect did not appreciably affect the majority-carrier mobility. However, interelectronic collisions modified the minority-carrier mobility by a factor of 3. The effect of carrier-carrier scattering has also been discussed by Yeoman and Young [139], who proposed that such processes dominate basal conductivity at high temperatures.

Carrier-carrier effects are normally treated by using a screened Coulomb potential in the approximation of Born scattering. Such scattering is therefore similar to that from ionized impurities, for which the relaxation time is usually written [51] as

$$\tau_I = \frac{K^2 (2m^*)^{1/2}}{\pi e^4 N_I} \frac{\epsilon^{3/2}}{g(n, T, \epsilon)} , \tag{117}$$

where K is the dielectric constant, N_I is the density of ionized impurities, and $g(n, T, \epsilon)$ is a slowly varying function of the density of carriers, temperature, and electron energy. The scattering probability is thus inversely proportional to the cube of the velocity. For electrons near the Fermi surface low-angle scattering predominates.

A detailed calculation of the relaxation time for scattering from ionized impurities in graphite has recently been reported by Boardman and Graham [59]. An earlier paper by McClure and Spry [140] used a screened self-consistent potential for the scattering centers based on the Thomas-Fermi model. Boardman and Graham showed that the anisotropy of the bands introduces important corrections to the dielectric constant describing the screening. Their calculated value for the residual resistivity of graphite with 10^{18} cm^{-3} ionized impurities was 1.1×10^{-5} ohm m, higher by a factor of 7 than the value calculated by using the Thomas-Fermi density-of-states model.

This calculation is particularly interesting for the interpretation of the effects of collision broadening on the De Haas-Van Alphen oscillations. The Dingle temperature $\Delta T = h/\pi k \tau_{coll}$ describes the effective change in temperature due to collisions with relaxation time τ_{coll}. This relaxation time is normally taken to be a constant, whereas strictly speaking it oscillates with the magnetic field. This implies that the relaxation time obtained by

fitting the De Haas-Van Alphen type of oscillations may differ by a factor of
3 or more from the value obtained from the mobility [84].

The results obtained (Table 3) are in fair agreement. Dingle temperatures obtained in irradiated material sensibly increase with acceptor level
[22].

Materials prepared from pyrolytic deposits are characterized by regions
of nearly perfect material separated by grain boundaries. Crystal regions
on either side of the boundaries are rotated with respect to each other and
may be tilted by a very small angle [97]. The resulting relaxation time for
electron scattering at the boundaries then depends largely on the basal
velocity v of the carriers and the mean crystallite dimension ℓ (Fig. 34):

$$\tau(\epsilon) \propto \frac{\ell}{v} \propto \ell \epsilon^p, \tag{118}$$

where $p = -1/2$ for parabolic bands. It has been established that the mean
basal crystallite dimension obtained from electron-microscope studies
agrees quite well with the value obtained from conductivity measurements
[8, 141], amounting to more than 1 μm (10^{-4} cm) in the best available material (Fig. 35).

In conclusion, it is realistic to assume that the electron-phonon interaction is predominantly responsible for the current-limiting collisions at
high temperatures (T $\geq$ 100°K) in graphite. In natural crystals ionized-
impurity scattering predominates at low temperatures, whereas in synthetic

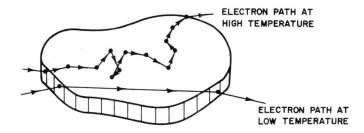

FIG. 34. Schematic representation of the scattering of carriers from
prismatic faults in graphite. At low temperatures the crystallite dimension
controls the mean free path. It is assumed that the carriers have a negligibly
small velocity component in the c-axis direction.

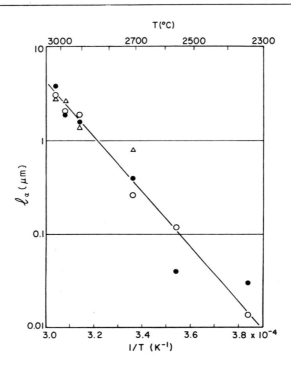

FIG. 35. Plot of the basal crystallite dimension for pyrolitic graphite deposited at various temperatures (o, X-ray data; •, mobility data at low temperatures; Δ , thermal conductivity data). The agreement between the various
techniques shows that electrons and phonons are scattered predominantly by
prism edges of crystallites at low temperatures. From Ref. [141].

---

materials scattering at grain boundaries predominates. At all temperatures
interelectronic collisions are important and modify other scattering mechanisms. The role of mutual-drag effects is not fully understood at present and
needs to be examined in detail. Mechanisms controlling c-axis conductivity
are discussed in Section VIII. A.

D. General Solutions to the Boltzmann Transport Equation

Once the details of the bands and the relaxation time are known, solutions for the conductivity coefficients can be obtained. If the quantity $\omega_c \tau$

is independent of energy, the denominator can be taken outside the conductivity integral to give a particularly simple form for the conductivity components. For contributions from N groups of carriers i in a magnetic field H parallel to the c axis (z coordinate axis)

$$\sigma_{xx} = \sum_{i=1}^{N} \frac{\sigma_{0i}}{1 + (H/H_i)^2} , \tag{119a}$$

$$\sigma_{xy} = \sum_{i=1}^{N} \frac{a_i n_i e_i cH/H_i^2}{1 + (H/H_i)^2} , \tag{119b}$$

σ_{0i} = zero-field conductivity of carrier i

$$= n_i e \mu_i = n_i ec/H_i,$$

where (120)

$$H/H_i \equiv \omega_i \tau_i, \tag{121}$$

e_i is the charge of carrier i, and c is the velocity of light in vacuum if units are cgs units.

These formulas were first used by Soule [87] and McClure [142] to analyze galvanomagnetic effects in graphite. The correction factor a_i (approximately unity) essentially implies that the Hall and conductivity mobilities are not equal and must be evaluated for each particular case. The formulas are fairly good approximations even when $\omega_c \tau$ varies with energy, in which case $H_i = H/\omega_c \tau$ is a weighted value.

McClure [142] has also obtained the following relationships for the total number of carriers (electrons plus holes):

$$n + p = \frac{2}{\pi ec} \int_0^\infty \sigma_{xx}(H) \, dH. \tag{122}$$

Similarly the difference in the zero-field conductivities of holes and electrons can be obtained from the following:

$$\sigma_0^{el} - \sigma_0^{h} = -\frac{2}{\pi} \int_0^\infty \frac{\sigma_{xy}}{H} \, dH. \tag{123}$$

High-field data are needed to calculate the integrals exactly.

In pure graphite the majority of the contribution to σ_{xx} comes from the majority holes and electrons with nearly equal densities. In this case the average mobility ($<\mu> = \sqrt{\mu_1\mu_2}$) can be calculated from

$$<\mu> \simeq \left(\frac{\Delta\rho}{\rho_0 H^2} c^2\right)^{1/2} \quad \text{(cgs units)}, \tag{124}$$

$$<\mu> \simeq \left(\frac{\Delta\rho}{\rho_0 H^2}\right)^{1/2} \times 10^8 \quad (cm^2/\text{V-sec}). \quad \text{(H in gauss)}. \tag{125}$$

Soule [87] and McClure [142] used the conductivity-tensor formulation to analyze galvanomagnetic data for natural crystals of graphite, and Spain et al. [7, 8] used it for synthetic graphite at low fields.

General expressions similar to Eqs. (119a) and (119b) can be obtained for the coefficients M_{xx}, M_{xy}, L_{xx}, L_{xy}, N_{xx}, N_{xy} [24]. For M_{xx} and M_{xy} they are (Table 13) as follows:

$$M_{xx} = \sum_{i=1}^{N} \frac{n_i e_i ck\beta_i \xi_i}{|e|H_i[1 + (H/H_i)^2]}, \tag{126}$$

$$M_{xy} = \sum_{i=1}^{N} \frac{n_i ck\gamma_i \xi_i}{H_i^2[1 + (H/H_i)^2]}, \tag{127}$$

where

$$\xi_i = \frac{\eta i}{kT}, \tag{128}$$

where η_i is measured relative to the band edge. The parameters β_i, γ_i are to be evaluated for the band model chosen. Specific values are given in Table 14 for the cylindrical and spheroidal band models. In the case of extreme degeneracy ($\xi \to \infty$) $\xi_i\beta_i$ and $\xi_i\gamma_i$ tend to constant values (Table 14).

It is to be noted that M_{xx} depends on the sign of the carrier, but M_{xy} does not. In some cases the factors β_i and γ_i can be positive, but in graphite M_{xx} is normally positive for holes, negative for electrons, and M_{xy} is negative for both carriers, so that electrons are to be associated with a negative thermoelectric power (see Section VII.I).

TABLE 13

Simplified Expressions for the Conductivity-Tensor Coefficients[a]

Tensor component	General multiband expression	Zero-field coefficient	Alternative expression for zero-field coefficient
σ_{xx}	$\displaystyle\sum_{i=1}^{N} \frac{\sigma_{xx}^{(i)}(0)}{1+(H/H_i)^2}$	$n_i e \mu_i$	$n_i |e| c/H_i$
σ_{xy}	$\displaystyle\sum_{i=1}^{N} \frac{\sigma_{xy}^{(i)}(0)}{1+(H/H_i)^2}$	$\dfrac{n_i e \mu_i^2}{c}\alpha_i H$	$\dfrac{n_i e\,[\mu_H^{(i)}]^2}{c}H$
M_{xx}	$\displaystyle\sum_{i=1}^{N} \frac{M_{xx}^{(i)}(0)}{1+(H/H_i)^2}$	$\dfrac{n_i e \mu_i k}{|e|}\beta_i$	$\sigma_{xx}^{(i)}\dfrac{e_i}{|e|}k\beta_i$
M_{xy}	$\displaystyle\sum_{i=1}^{N} \frac{M_{xy}^{(i)}(0)}{1+(H/H_i)^2}$	$\dfrac{n_i \mu_i^2 k}{c}\gamma_i H$	$\dfrac{|\sigma_{xy}^i|}{|e|}k\gamma_i$

[a] μ_i = conductivity mobility = c/H_i.

H_i = characteristic field of carriers, related to the conductivity mobility. In this treatment H_i takes the same value for all tensor components.

$\mu_H^{(i)}$ = Hall mobility.

i = 1, 2,..,N = carrier index.

c = velocity of light.

α_i, β_i, γ_i are factors that can be calculated from the Boltzmann transport equation.

TABLE 14

Expressions for the Galvanomagnetic and Thermomagnetic Parameters $\bar{n}$, μ, α, β, γ for the Cylindrical and Spheroidal Band Models[a]

Parameter	Cylindrical model		Spheroidal model	
	p = 0	p = 1/2	p = -1/4	p = -1/2
n	$\dfrac{m^*kT}{\pi C_0 h^2} F_0(\xi)$ $\left[\dfrac{m^*kT}{\pi C_0 \hbar^2}\right]\zeta$	Same as for p = 0. Same as for p = 0	$\dfrac{1}{2\pi^2}\dfrac{\sqrt{8m_3}}{\hbar^3}m_1(kT)^{3/2} F_{1/2}(\xi)$ $\left[\dfrac{2}{9\pi^2}\dfrac{\sqrt{8m_3}}{\hbar^3}m_1\eta\right]^{3/2}$	Same as for p = -1/4 Same as for p = 1/4
μ	$\dfrac{|e|\tau(T)}{m^*}$	$\dfrac{|e|\tau(T)}{m^*}$	$\dfrac{e\tau_0}{m_1}\dfrac{5}{4}\dfrac{F_{1/4}}{F_{1/2}}(kT)^{-1/4}$ $\left[\dfrac{3}{2}\dfrac{e\tau_0}{m_1}\eta^{-1/4}\right]$	$\dfrac{e\tau_0}{m_1}\dfrac{F_0(\xi)}{F_{1/2}(\xi)}(kT)^{-1/2}$ $\left[\dfrac{3}{2}\dfrac{e\tau_0}{m_1}\eta^{-1/2}\right]$
α	1 [1]	1 [1]	$\dfrac{16}{25}\dfrac{F_{1/2}(\xi) F_0(\xi)}{F_{1/4}^2(\xi)}$ [2/3]	$\dfrac{2F_0^2(\xi)}{F_{-1/2}(\xi) F_{1/2}(\xi)}$ [3/4]

100

β	$\xi - 2\dfrac{F_1(\xi)}{F_0(\xi)}$	$\xi - \dfrac{5}{3}\dfrac{F_{3/2}(\xi)}{F_{1/2}(\xi)}$	$\xi - \dfrac{9}{5}\dfrac{F_{5/4}(\xi)}{F_{1/4}(\xi)}$	$\xi - \dfrac{2F_1}{F_0}$
	$\left[\dfrac{\pi^2}{3\xi}\right]$	$\left[\dfrac{\pi^2}{2\xi}\right]$	$\left[\dfrac{5\pi^2}{12\xi}\right]$	$\left[\dfrac{\pi^2}{3\xi}\right]$
γ	$\xi - 2\dfrac{F_1}{F_0}$	$\xi - \dfrac{5}{3}\dfrac{F_{3/2}(\xi)}{F_{1/2}(\xi)}$	$\dfrac{16}{25}\dfrac{F_{1/2}(\xi)}{F_{1/4}(\xi)}[\xi F_0(\xi) - 2F_1(\xi)]$	$\dfrac{F_{1/2}(\xi)}{2F_0^2(\xi)}[\xi F_{-1/2}(\xi) - 3F_{1/2}(\xi)]$
	$\left[\dfrac{\pi^2}{3\xi}\right]$	$\left[\dfrac{\pi^2}{2\xi}\right]$	$\left[\dfrac{2\pi^2}{9\xi}\right]$	$\left[\dfrac{\pi^2}{9\xi}\right]$

[a]Here ξ is the reduced electrochemical potential $\equiv \eta/kT$; values in brackets refer to the case of extreme degeneracy.

101

E. Basal Conductivity in Zero Magnetic Field

Measurements of the basal conductivity as a function of temperature for natural and synthetic specimens are shown in Fig. 36. Several authors [25, 77, 105, 128, 143] have fitted these data to within a few percent using various models for the energy bands and relaxation times. One feature is common to all of these calculations. The number of carriers is fitted to data at low temperatures and then calculated at higher temperatures, giving a result that is much higher than that found experimentally (see Fig. 44). In order for the conductivity to be fitted, a mobility lower than that found experimentally is calculated. For this reason the number of carriers and their

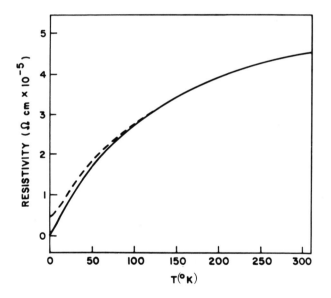

FIG. 36. The basal resistivity of graphite as a function of temperature for a natural crystal (solid curve) [87] and a high-quality synthetic crystal (broken curve) [8]. The results are normalized at 300°K for purposes of comparison. (Actual differences at 300°K are greater than the expected error of measurement.) The main differences in behavior occur at a very low temperature as a result of scattering from prismatic faults in synthetic crystals.

mobilities are treated separately here. In general it would seem better to fit these quantities than the conductivity.

The effect of pressure on the basal conductivity has been measured by several authors (Table 15). At $300^{\circ}K$ the pressure coefficient of resistance is very small ($\sim 10^{-6}$ bar^{-1}). Measurements of the magnetoresistance indicate that this arises fortuitously since the increase in the number of carriers is almost exactly compensated by the decrease in the mobility [102, 146]. A sketch of the variation of the pressure coefficient of resistance with temperature is shown in Fig. 37, and separate discussions of the mobility and density dependence are given in the following sections.

F. Magnetoresistance and Mobility of the Charge Carriers

The average mobility of the charge carriers can be obtained immediately from Eqs. (124) and (125). This relationship implies that the resistance

TABLE 15

Effect of Pressure on Resistance of Graphite[a]

Material	Pressure range (bars)	Temperature ($^{\circ}K$)	$\dfrac{d \log \sigma_{11}}{dp}$ (M bar^{-1})	$\dfrac{d \log \sigma_{33}}{dp}$ (M bar^{-1})	Ref.
Natural flake (Ceylon)	0–100,000	Room	+5	--	144
Pyrolytic	0–400,000	Room	--	--	145
Natural flake[b]	0–600,000	77 and 296			27
Natural flake (Ceylon and Ukraine)	0–9000	290–420	+3.0	--	146
Pyrolytic	0–8000	4.2	--	--	100
Hot-pressed and annealed graphite	0–400	4.2 77 273	28 12–14 ±1	28.2 ± 1 27.6 ± 1 25.5 ± 1.5	139 139 139
Hot-pressed pyrolytic	0–4000	77 300	10–13 ±1		102 102

[a]It should be noted that some authors report nonlinear effects below 50 bars.
[b]Cubic phase of graphite reported.

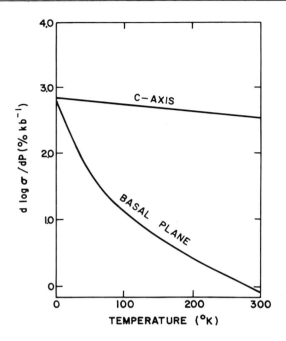

FIG. 37. Variation in the pressure coefficient of resistance for current flow both parallel and perpendicular to the basal planes. Adapted from Ref. [139].

varies as the square of the magnetic field. It is well established that in both natural crystals [87, 147] and synthetic materials [50, 148] the exponent in field varies with temperature and field strengths. For moderate fields ($H \doteq H_i$) the exponent has a value of approximately 1.85. Unusual variations in very low fields can probably be attributed to minority-carrier effects. Since the exponent is less than 2, mobility values calculated as above become smaller as the field increases. Soule [87] for measurement convenience adopted a standard of measurement at 3 kG, and other workers have in general followed this practice. [A better standard would be to choose values taken at a field related to the characteristic field of the carriers (e.g., H_i or $0.5H_i$).]

Figure 38 shows the variation in mean mobility with temperature for natural and synthetic specimens. Below 300°K the exponent in temperature of the mobility variation is approximately -1.2. Less perfect specimens have a less steep slope. Klein [50] found that the slope increased to approximately -1.5 for temperatures above 300°K. The results are not at variance with a model in which electron-phonon scattering predominates. At very low temperatures the curves bend over to attain a maximum mobility controlled by scattering at defects. For the crystal EP-11 [87] the very high mobility at 4.2°K (2×10^{6} cm^{2}/V-sec) corresponds to a relaxation time of approximately 5×10^{-11} sec and a mean free path ℓ of about 10^{-3} cm (10 μm).

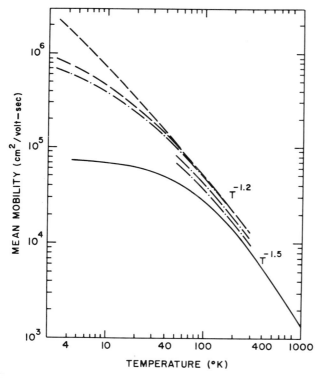

FIG. 38. Variation in mean mobility for the carriers along the basal planes in graphite. Values are calculated from the magnetoresistance as explained in the text. Key: ---, Soule [87]; -·-, Spain et al. [7]; ——, Klein [50].

When the magnetic field is tilted away from the c axis, the magnetoresistance decreases, until a minimum value is reached for the magnetic
field parallel to the planes [8]. Although the anisotropy factor is greater
than 200 for some specimens, the residual magnetoresistance in part is a
satellite effect of the principal transverse magnetoresistance arising from
small crystallite misorientations [97]. The angular dependence of the

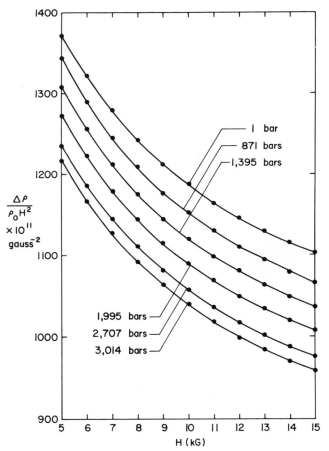

FIG. 39. Variation with magnetic field and hydrostatic pressure of the
magnetoresistance parameter $\Lambda\rho/\rho_0 H^2$, which is proportional to the square
of a mean mobility value. The curves indicate clearly how the value of the
mobility calculated from magnetoresistance data depends strongly on the
field strength (T = 296°K). From Ref. [102].

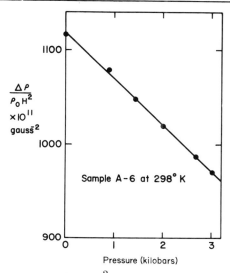

FIG. 40.   Variation of $\Delta\rho/\rho_0 H^2$ with pressure, showing that the quantity $(<\mu>)^2$ varies linearly with pressure.   Measurement at 298°K.   This result in consistent with a model in which electron-phonon scattering predominates. From Ref. [102].

magnetoresistance is not a classical $\cos^2\theta$ dependence but may be fitted accurately as a principal magnetoresistance effect with magnetic-field component $H\cos\theta$ parallel to the c axis.

In general the mobility values calculated by using the approximate formulas (124) and (125) do not differ by more than a few percent from the mean value for the majority carriers obtained by a complete analysis of the magnetoconductivity data.   The mobility calculated in this way is a very convenient measure of specimen perfection, although an incomplete description of the defect characteristics of a material is obtained in this way [8]. The effect of pressure on the magnetoresistance has been measured between 290 and 420°K by Arkhipov et al. [76], Likhter and Kechin [146], and at 290 to 300°K by Spain [102].   The linear decrease in the parameter $\Delta\rho/\rho_0 H^2$ (which is proportional to the square of the mean mobility) is large, amounting to approximately 4% per kilobar (Figs. 39 and 40) and is consistent with a model of the relaxation time in which electron-phonon scattering predominates.   At 77°K the effects are more complex, and have not been analyzed satisfactorily [102].   However, both at this temperature and at 4.2°K [100] the magnetoresistance is reduced with application of pressure.

At low temperatures quantum oscillations are observed (Shubnikov-de Haas oscillations) (Fig. 41), as discussed in Section VI-C (see also Table 3).

The magnetoresistance has been measured in pulsed magnetic fields of up to 160 kG by McClure and Spry [140], who found that the diagonal component of the magnetoconductivity tensor (σ_{xx}) varied as the inverse of the magnetic field, whereas the magnetoresistance was approximately linear. Using a screened potential in the Thomas-Fermi approximation for ionized-impurity scattering and allowing for the changes in screening length with magnetic field, satisfactory agreement with the experiment was obtained.

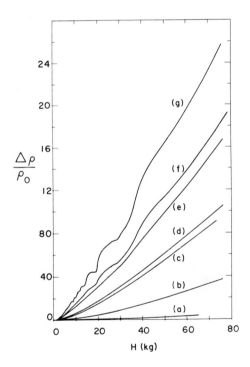

FIG. 41. Variation of the magnetoresistance with magnetic field at several temperatures (°K): (a) 307, (b) 77, (c) 46, (d) 31, (e) 20, (f) 4.2, (g) 2. The onset of Shubnikov-de Haas oscillations at low temperatures is evident. Techniques that measure the slope of the curves reveal a wealth of detail concerning these oscillations not revealed in this diagram. Ref. [149].

A more recent measurement by Woollam [108] in steady fields to 150 kG
showed that the behavior was not linear. Refinements to the calculation [48]
show that the linear behavior is not to be expected over the entire field range.

An interesting effect, known as the Esaki kink effect, has recently been
observed in graphite. In a magnetic field, when the drift velocity of the carriers equals the velocity of propagation of the lattice modes, a kink appears in
the voltage-current relationship [150, 151] (Fig. 42). More recent measurements of Yugo, Hayashi, and Yazawa [152] showed that at least three kinks
could be seen in the curves (Fig. 43) corresponding respectively to interaction with the in-plane transverse mode ($s_1 = 1.23 \times 10^6$ cm/sec), in-plane
longitudinal mode ($s_2 = 2.10 \times 10^6$ cm/sec), and out-of-plane mode traveling
along the basal planes ($s_3 \simeq 3.6 \times 10^5$ cm/sec). Reasonable agreement was
obtained between calculated drift velocities and wavepropagation velocities,
except for the latter mode, for which a value of $s_3 = 6.3 \times 10^5$ cm/sec was
obtained. This effect clearly illustrates the fact that the electrons interact
with the transverse modes as discussed in Section VII-C.

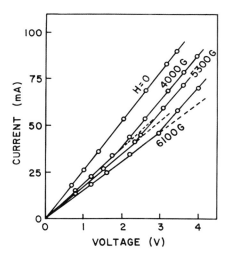

FIG. 42. Illustration of the Esaki-kink effect in graphite. At the point
where the slope of the current-voltage characteristic changes, the mean drift
velocity of the carriers equals the velocity of propagation of the interacting
phonon mode. Adapted from Ref. [151].

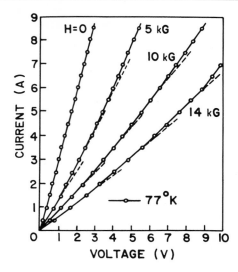

FIG. 43. Illustration of the multiple-kink effect in graphite in which the average drift velocity of the carriers successively exceeds the wave-propagation velocity of three different modes of vibration. Ref. [152].

G. Number of Carriers

An approximate value for the total number of carriers can be obtained from a measurement of the zero-field resistivity and the magnetoresistance, using relationships (120) and (124). Alternatively a complete magneto-conductivity analysis may give more accurate figures that do not differ from the approximate estimates by more than a few percent. Values obtained for natural and synthetic crystals are given in Table 16 and are plotted in Fig. 44. Theoretical calculations of Ono and Sugihara [42] and Kechin [25] are in-cluded for comparison, giving a higher value at room temperature than ex-periments indicate (see also Table 17). This discrepancy is not removed by taking into account the effect of thermal expansion on the band parameters. The c-axis spacing increases by 0.5% in the temperature interval 0 to 300°K [153], and a pressure of 1 kilobar reduces the spacing by 0.25% [154]. The change in the number of carriers can be estimated from the pressure coef-ficient of the conductivity and magnetoresistance at room temperature (see Sections VII. E and VII. F) to amount to no more than a few percent.

TABLE 16

Total Carrier Concentrations, Average Mobilities, and Majority-Carrier
Mobility Ratios Obtained from Galvanomagnetic Data

Sample	T (°K)	n^d (cm^{-3} x 10^{18})	$\bar{\mu}$ (cm^2/V-sec) x 10^4	b^a	Type of minority carrier
EP-14[b]	300	9.9	1.4	1.10	Electron
	77	3.3	6.8	0.87	Hole
	4.2	5.2	70	0.79	Electron
SA-26[c]	300	11.3	1.25	1.09	Electron
	77	4.2	5.9	0.95	Electron
	53	3.5	10.0	0.97	Electron
	4.2	3.0	48	1.11	Electron

[a] $b = \mu_1/\mu_2$.

[b] Data from Ref. [87]. Natural single crystal.

[c] Data from Ref. [8]. Hot pressed and annealed specimen.

[d] Calculated from $n = \sigma/e\,\bar{\mu}$

TABLE 17

Carrier Properties Deduced from Multicarrier
Analysis of Galvanomagnetic Data[a]

Sample	T (°K)	Majority electron		Majority hole		Minority electron		Minority hole	
		n_1 x10^{-18}	1 x10^{-4}	n_2 x10^{-18}	2 x10^{-4}	n_3 x10^{-16}	3 x10^{-4}	n_4 x10^{-16}	4 x10^{-4}
EP-7[b]	300	7.36	1.09	7.22	0.99	--	--	0.057	150
	77	2.46	4.86	2.39	6.53	--	--	0.87	200
	2	2.14	59.1	2.08	64.5	20.0[c]	0.7	--	--
EP-14[b]	300	7.04	1.13	7.04	1.01	0.05	39	--	--
	77	2.24	6.38	2.19	7.33	--	--	0.33	57
	2	2.92	83.9	2.88	104	20[c]	0.7	--	--
SA-20[d]	77	2.25	5.07	2.25	5.05	3.3	20.9	--	--
SA-26[d]	77	1.98	5.82	1.98	6.20	2.44	22	--	--

[a] Carrier density expressed in units of cm^{-3}; mobility expressed in units of cm^2/V-sec.

[b] Data from Refs. [87] and [142]. Natural single crystal.

[c] This carrier is not related to the low-field behavior of the Hall coefficient. Data only fitted in region above 250 gauss.

[d] Data from Ref. [8]. Hot pressed and annealed graphite.

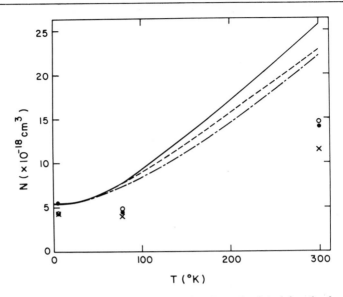

FIG. 44. Variation in free-carrier density calculated for the four-parameter band model with parabolic [25] and hyperbolic [77, 105] energy-wavevector relationships at low temperatures. Agreement with galvanomagnetic data is good at 0°K, but at high temperatures the experimental values lie well below the theoretical curves. Experimental data indicate a slight reduction in the number of carriers at 77°K compared with 4.2°K. Theoretical curves:___·___, Ono and Sugihara [77];___, McClure and Smith [105]; ----, Kechin [25]. Experimental values: ● , Soule [87], sample EP-7; o, Soule [87], sample EP-14; x, Spain et al. [8], sample SA-26.

Essentially the change in the number of carriers is overestimated by the calculations, implying that the thermal energy kT is overestimated compared with the electrochemical potential η. This could be corrected by using a larger value for $|\gamma_2|$ requiring a smaller value of $|\gamma_1|$ to match the data at low temperatures [see Eq. (62)]. Alternatively the results could be explained if the band parameters depend on the temperature. Since the number of carriers is a parameter of such fundamental importance, it is surprising that the discrepancy mentioned here has not previously been discussed.

The effect of pressure on the number of free carriers can be estimated from galvanomagnetic data [76, 100, 102, 146]. There is a little difference between the values obtained at 4.2 and above $300°K$ (d log n/dp $\approx$ 3-4 x 10^{-2} kilobar $^{-1}$). This value also agrees with the estimates found from the effect of pressure on the De- Haas-Van Alphen oscillations [91, 92]. A summary of the dependence of the band-overlap parameters on pressure is given in Table 4.

H. Hall Effect

Since the number and mobility of the majority carriers are comparable in graphite, the Hall coefficient is small ($\sigma_{xy} \ll \sigma_{xx}$) and has a remarkable variation with temperature and magnetic field [147]. Spain, Ubbelohde, and Young [8] have shown that for small magnetic fields (H $\leq$ 5 kG) the Hall coefficient of the highest quality synthetic material closely approximated that of natural crystals [87], except for differences in minority carrier behavior (Figs. 45 and 46). More recent measurements at higher fields [102, 149] suggest that differences may also exist in this field range (Figs. 47 and 48).

A comparison of data obtained from magnetoconductivity analysis is given in Table 16. The variation in the mobility ratio b = $\mu_{e\ell}/\mu_h$ for these data is indicated in Fig. 49. Trends in the behavior of synthetic crystals can largely be identified with changes in crystallite size, affecting the scattering length at low temperatures. The data are readily explained if γ_2 is taken as negative [23]. This conclusion is based mainly on the following points:

1. The average hole mobility is greater than the electron mobility below about $200°K$ in the most perfect specimens, favoring a heavy-electron model.

2. In a specimen with boundary scattering predominating at low temperatures the ratio b is increased at low temperatures, consistent with a mobile-hole model.

3. At a given temperature the ratio b for different specimens decreases as the crystallite scattering length increases, favoring a mobile-hole model.

Differences in the properties of the mobile minority carriers in natural and synthetic specimens have been less easily explained. The mobile hole observed in natural crystals at $77°K$ (giving the sharp upward turn in the Hall curves at low fields) are consistent with the new model, with Δ positive and γ_2 negative. They can be identified with the minority-carrier pocket at point H. Using the extremal cross sections obtained by Soule, McClure, and

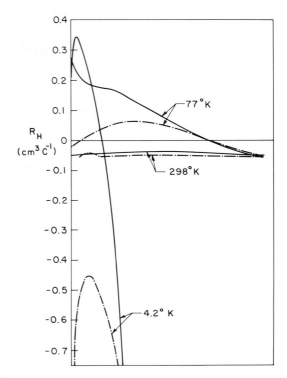

FIG. 45. Variation of the Hall constant with magnetic field for a natural crystal of graphite (EP-7, solid curves [87] and a high-quality synthetic crystal (SA-26, broken curves [8] at 298, 77, and 4.2°K. At very low fields the natural crystal indicates the presence of a mobile minority electron at 4.2 and 298°K, but a hole carrier at 77°K. The result for the synthetic crystal is consistent with a mobile minority electron at all three temperatures.

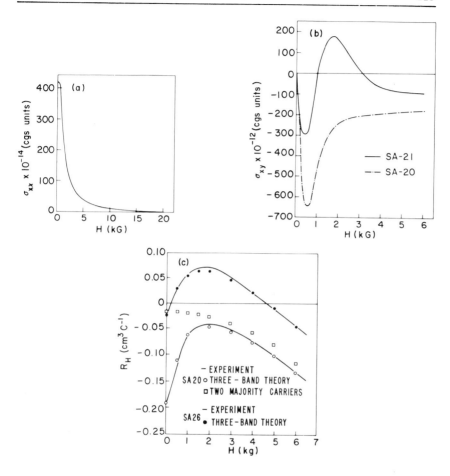

FIG. 46. (a) The diagonal component σ_{xx} (H_z) of the magnetoconductivity tensor for a natural crystal at 77°K [87]. (b) The off-diagonal component σ_{xy} (H_z) of the magnetoconductivity tensor for two synthetic specimens of graphite at 77°K [7]. Regions of positive σ_{xy}(H) correspond to regions of positive Hall coefficient. (c) A sketch illustrating the fit that can be obtained to Hall data at 77°K for fields less than about four times the characteristic fields of the majority carriers using a three-band model (majority electron, majority hole, and mobile minority electron). For comparison the Hall curve for only the two majority carriers is also indicated. From Ref. [7].

Smith [84] and assuming a spheroidal shape, the pocket should contain approximately 3 x 10^{15} carriers per cubic centimeter at low temperatures,
in agreement with the Hall data [87]. In high-quality synthetic materials the
downward turn in the Hall curves at low fields can be identified with a mobile
minority electron (Table 16). The minority pocket observed by Williamson,
Foner, and Dresselhaus [5] in similar material contains 3 x 10^{15} carriers
per cubic centimeter compared with a value of 3 x 10^{16} observed in the Hall
effect. It is probably safe to conclude that the minority electrons are not

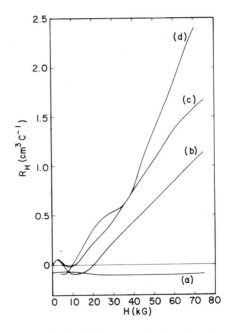

FIG. 47. Variation in the Hall constant for a synthetic crystal of graphite for fields of up to 80 kG (8 teslas) at (a) 307°K, (b) 77°K, (c) 46°K,
(d) 20°K. The onset of Shubnikov-De Haas oscillations can be clearly seen.
The further crossing of the Hall curve at T = 77°K, H = 25 kG may result
from quantum effects. (Data from Ref. [149].)

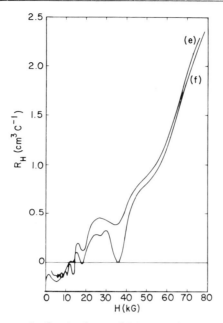

FIG. 48.   Curves similar to those of Fig. 47 plotted to the same scale at lower temperatures: (a) $4.2°K$; (b) $2°K$. Data from Ref. [149].

located near point H. This is supported by the very small effect of pressure on the low-field Hall curves (Fig. 50) [102]. Analysis indicates that the number of minority electrons varies by less than 1% per kilobar. This compares with a variation of 6% per kilobar for the extremal area [92].

More recently, Cooper, Woore, and Young [155] have found that the basal crystallite dimension of hot-pressed and annealed graphite can be systematically increased by further annealing. The subsequent behavior of the low-field Hall constant then bridges the gap between previous results on synthetic crystals at $77°K$ and natural crystals (Fig. 51). However, the variation in the Hall constant for natural crystals at $4.2°K$ indicates that the minority carrier is an <u>electron</u> (Fig. 45), so that the results are still not completely understood at present.

Recent measurements [149] of the Hall constant in synthetic crystals to 80 kG confirm that a further crossover occurs at about 30 kG [102] (Figs. 47 and 48). This implies that a three-band model is not sufficient to account for the observed variation. There is no satisfactory quantitative account for

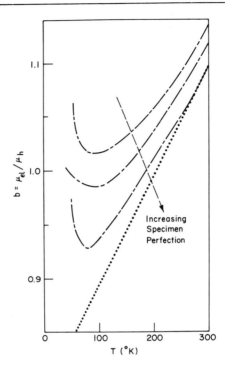

FIG. 49. Variation of the majority-carrier ratio b = $\mu_{e\ell}/\mu_h$ with temperature for a natural crystal (....) [87] and three synthetic specimens (_._) [7]. The trend is to lower values of b for specimens of higher perfection, in agreement with the assignment of electrons near point K.

this behavior at present, but it is possible that quantum effects are responsible.

Direct theoretical calculations of the Hall coefficient have not in general fitted the data well. The papers by Ono and Sugihara (γ_3 = 0 [77]; γ_3 = 0.20 eV [42]) treated the bands by using the nonparabolic dispersion relationships (11) and (12) but obtained solutions to the Boltzmann transport equation in the effective-mass or parabolic approximation. The error introduced in this way is difficult to assess. The authors pointed out [77] that the data could best be fitted with a negative value of γ_2. A detailed calculation has been made by Kechin [25] in the approximation of parabolic energy-wavevector relationship in the basal plane. Reasonable agreement with the zero-field

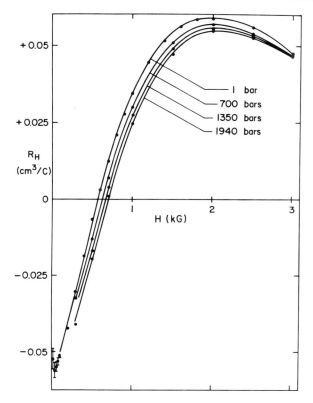

FIG. 50. Detail of the low-field Hall coefficient of graphite at 77°K for a high-quality synthetic crystal (sample A-6). Hydrostatic pressure changes the properties of the carriers by less than 1% per kilobar. It is possible that at very low fields the Hall coefficient becomes more positive, indicating the presence of a mobile minority hole in addition to the electron. Data from Ref. [102].

Hall coefficient was obtained. It is interesting to examine the numerical equation given for the variation of the Hall constant with temperature applicable for $T \geq 50°K$:

$$R(0) = \frac{0.487 a_0^2 C_0 \rho_T^2}{-|e|(kT)^2} \frac{\gamma_0^2 \gamma_2}{|\gamma_1|} \left(1 - 17.02 \left|\frac{\gamma_4 kT}{\gamma_0 \gamma_2}\right| - 1.247 \left|\frac{\gamma_4 \gamma_2}{\gamma_0 kT}\right|\right), (129)$$

where ρ_T is a parameter approximately equal to unity.

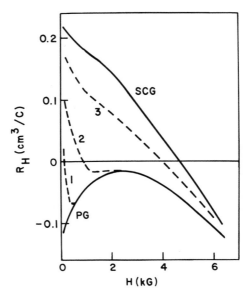

FIG. 51.   The change in the low-field behavior of the Hall constant of graphite at 77°K occasioned by an increase in crystallite size by annealing procedures.   From Ref. [155].

Equation (129) shows that at low temperatures the sign of the Hall coefficient is determined by $-\gamma_2/|e|$ that is, by $-\gamma_2$.   Since the measurements indicated a positive value, Kechin independently proposed that γ_2 should be negative.   In order to fit the data, γ_4 was chosen as positive.   Both of these assignments are in agreement with those of Schroeder et al. [20].   The sensitivity of Eq. (129) to γ_4 comes from the fact that this parameter alters the density of states and controls the variation in electrochemical potential with temperature (section V).

The agreement between experiment and Kechin's theory is in part fortuitous.   First, the parabolic model does not describe the carriers near the zone corners correctly, and the sign of the Hall coefficient depends sensitively on these carriers.   Second, the hyperbolic nature of the bands itself brings about a change in the electrochemical potential with temperature.   Third, the form of the relaxation time used neither includes the energy dependence nor considers changes in scattering brought about by the c-axis modes.   It is also noted that the zero-field Hall coefficient of itself is not a

particularly useful quantity. A complete description of the Hall coefficient
needs also to take into account the magnetic field dependence.

Formula (129) may be useful as a first approximation for analyzing the
pressure derivatives of the Hall constant, particularly at higher temperatures
($>300^{\circ}$K) for which experimental results have been reported [75, 102, 146].
Increase in pressure produces a linear change from negative values to more
positive ones (Fig. 52).

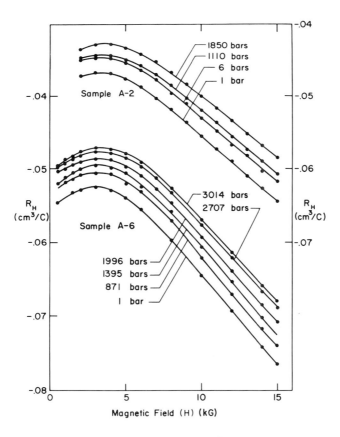

FIG. 52. The effect of pressure on the Hall constant of two synthetic
crystals of graphite near 300°K. Data from Ref. [102]. The left-hand scale
is for sample A-6, the right-hand scale for A-2, a less perfect specimen.

I.   Thermoelectric Power

The thermoelectric power S_{xx} in a magnetic field H_z can be obtained from

$$-S_{xx}(H) = \frac{\sigma_{xx} M_{xx} - \sigma_{xy} M_{xy}}{\sigma_{xx}^2 + \sigma_{xy}^2}.$$ (130)

In zero magnetic field σ_{xy} and M_{xy} are zero, so that

$$S_{xx}(0) = \frac{-M_{xx}(0)}{\sigma_{xx}(0)} = \frac{-\sum\limits_{i=1}^{N} M_{xx}^i(0)}{\sum\limits_{i=1}^{N} \sigma_{xx}^i(0)}.$$ (131)

The quantities $M_{xx}(0)$ and $\sigma_{xx}(0)$ can be calculated for specific cases. For the two-band cylindrical model the formulas are particularly simple.   If $\tau = \tau_0 \epsilon^p$, then the diffusion component of S is [Table 13 and 14].

$$S_{xx}^{el}(0) = \left[-\frac{p+2}{p+1} \frac{F_{p+1}(\eta)}{F_p(\eta)} + \xi \right] \frac{k}{e}$$ (132)

In the limit of extreme degeneracy ($\eta \gg kT$) this becomes, for $p \simeq o$,

$$S_{xx}^{el}(0) \simeq -\frac{p+1}{3} \frac{\pi^2 k^2 T}{e\eta},$$ (133)

and a similar expression holds for the holes.   It can be seen that for temperatures close to the degeneracy temperature ($kt \simeq \eta$) the contribution from the individual bands may be as high as 300 $\mu V/^\circ K$ (p = 0) while the measured thermoelectric power is usually less than 10 $\mu V/^\circ K$.   This compensation makes the absolute magnitude of the thermoelectric power very difficult to calculate.   In principle it would be better to fit theoretical calculations directly to M_{xx} and M_{xy}, but to date no complete data have been reported.

Representative data for the thermoelectric power as a function of temperature are plotted in Fig. 53.   The only measurement through the complete temperature range 7 to 300°K has been made on ironmelt crystals [156].   Results on less perfect synthetic material in the temperature range 60 to 300°K [8] are similar in form, though displaced to more negative values as the crystallite dimension is reduced.   The result of Blackman, Saunders, and

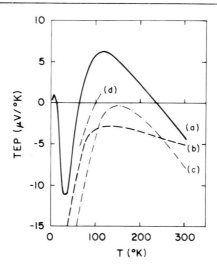

FIG. 53. Representative data of the variation in the thermoelectric power (TEP) of graphite with temperature. Data from Takezawa and Tzuzuku [156] for (a) iron-melt crystal; (b) pyrolytic crystal. Data from Spain et al. [8] for (c) hot-pressed pyrolytic graphite; (d) annealed, hot-pressed pyrolytic graphite.

Ubbelohde [148] obtained on recrystallized pyrolytic graphite above 80°K does not fit in with this general scheme except above 100°K, and the result of Tamarin, Shalyt, and Volga [157] on similar material below 100°K is somewhat ambiguous, since a discussion of the sign of the measured coefficient was not given.

Takezawa et al. [156] fitted their data to the simple expression

$$S_{diff} = \frac{k}{e} \frac{(a - b)}{(a + b)} \left(\frac{2F_1}{F_0} - \frac{\eta}{kT} \right), \tag{134}$$

which was first obtained by Klein [50]. This is the total thermoelectric power for the cylindrical band model with $p = 0$, $\eta = \epsilon_0/2 = \gamma_2$ [see Eqs. (32) and (33)]. Using experimental values for the galvanomagnetic ratios $a = n_h/n_{el}$, $b = \mu_{el}/\mu_h$, the diffusion term in the total thermoelectric power, S_{diff}, can be separated from the phonon-drag component (S_g (Fig. 54). Similar calculations by Spain et al. [8] indicate that the downward trend of the curves below about 120°K could be satisfactorily accounted for on the basis of the

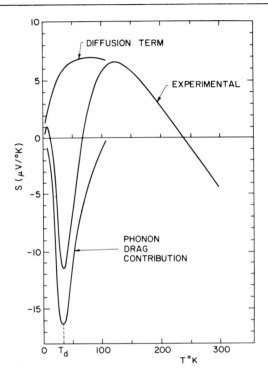

FIG. 54. The diffusion and phonon-drag components of the thermoelec-.
tric power [156].

diffusion thermopower alone. In this temperature range the galvanomagnetic
ratio b rises sharply with temperature in good-quality synthetic specimens
(see Section VII. H and Fig. 49), whereas the mobile minority electron may
contribute more than 5 μV/°K to the total thermoelectric power. Specifically,
it is difficult to assess accurately the contribution S_g unless S_{diff} is known
precisely (see also [105]).

Sugihara [158] has presented a theory to account for the phonon-drag
thermoelectric power in graphite, using the spheroidal band heavy-electron
model. This contribution arises at low temperatures, where the phonon-
phonon interaction is small [40], so that phonon momentum in the direction of
the temperature gradient can be imparted to the carriers. In the steady state,
an emf is set up, in addition to the normal thermoelectric effect that origin-
ates from the diffusion of the carriers down the gradient in the electrochemical

potential. Sugihara [158] found that when Normal electron-phonon scattering processes were taken into account (between the carriers and the in-plane longitudinal mode only), the resulting drag effect on each carrier gave a contribution to the thermoelectric power as high as 200 $\mu V/^\circ K$ near the de-generacy temperature, so that self-cancellation effects were again important ($S_g \simeq 13 \ \mu V/^\circ K$ experimentally). Using simple models of the electron-phonon interaction, the coupling between electrons and phonons should be greater than that between holes and electrons, thus producing a negative contribution to S_g, in agreement with the experimental values, if γ_2 is negative.

At very low temperatures both the diffusion and phonon-drag contributions to the thermoelectric power should tend to zero. In particular the phonon-drag component should be proportional to the lattice specific heat (αT^2 in graphite). At higher temperatures, where phonon-phonon scattering domin-ates the phonon mean free path, Sugihara [158] predicted that the term should be proportional to T^{-3}. This behavior qualitatively fits the results.

Jay-Gerin and Maynard [159] have also considered the phonon-drag ef-fect in graphite using the spheroidal band model. They conclude that the pre-dominant phonon-electron interaction arises from phonons with wavevector $q = 2\kappa_F$ leading to a maximum contribution to the phonon-drag at a temperature

$$T = \frac{\hbar\omega(q_F)}{k} \simeq 43.5^\circ K, \tag{135}$$

where $\hbar\omega$ (q_F) is the phonon energy associated with the mode q_F and k is Boltzmann's constant. Jay-Gerin and Maynard concluded that the sign of the contribution would only be correct (negative) if the parameter γ_2 were nega-tive. It was also concluded that trigonal warping of the surfaces did not change the magnitude of the effects appreciably.

Sugihara and co-workers [160] have also demonstrated that the phonon-drag contribution to the thermoelectric power is greatly enhanced in magnetic fields (Fig. 55). At higher temperatures the variation in thermoelectric power with magnetic field is quite complex, involving changes of sign similar to those observed in the Hall effect [24]. From the tensor component M_{xx} it is possible to obtain information about the minority carriers complementary to that obtained with the Hall coefficient.

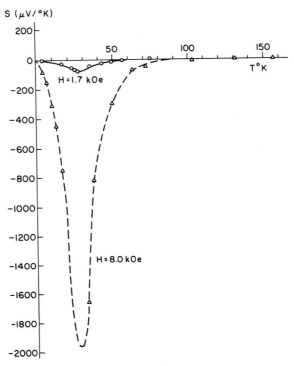

FIG. 55. The change in the thermoelectric power of graphite in a magnetic field, illustrating the enhancement of the phonon-drag effect, [160].

The formulas for the rmoelectric and thermomagnetic effects involve
integrals of the form [24]

$$I = \int f \text{ (relaxation time) } v_x^2 (\epsilon - \eta) \frac{\partial f_0}{\partial \epsilon} d\epsilon. \tag{136}$$

The factor $(\epsilon - \eta)$ ensures that the coefficients β and γ (Table 13) are a
measure of the asymmetry of the relaxation time and the density of states
about the electrochemical potential. It is for this reason that the calculation
of the thermoelectric properties is very sensitive to the exact model of the
bands and relaxation time used. It is probable that a semiempirical calculation represents the best that can be achieved for graphite.

J. Nernst Coefficient

The variation in the Nernst coefficient with temperature is indicated in Fig. 56. [157, 161]. Wright [162] and Goldsmid and Lacklison [163] have also discussed the Nernst effect in graphite in relation to Ettingshausen cooling devices.

Mills et al. [161] and Tamarin et al. [157] used an expression for the Nernst coefficient Q based on spherical bands of normal and inverted form. The rigorous expression for Q is (Table 10)

$$-Q = \frac{1}{H}\left(\frac{\sigma_{xy}M_{xx} - \sigma_{xx}M_{xy}}{\sigma_{xx}^2 + \sigma_{xy}^2}\right). \tag{137}$$

If suitable approximations are made, this expression gives the usual two-band formula with variables σ, R, S (see, for example, Ref. [164]). Using

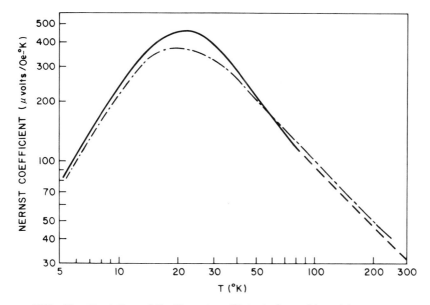

FIG. 56. Variation of the Nernst coefficient of graphite with temperature. The theoretical curve −·− [24] includes only the diffusion component of the coefficient for the spheroidal band model of graphite. Experimental data are from Mills et al. [161] (−−−−) and Tamarin et al. [157] (——).

experimental values for σ_{xx} and σ_{xy}, and computing values for M_{xx} and M_{xy} by using the four-parameter band model, good agreement with the experimental results can be obtained above 20°K [24]. The result obtained is not sensitive to the particular band model used. This is because the first term in the numerator is negligible compared with the second term (although Tamarin et al. [157] conclude differently) and because σ_{xx} and M_{xy} are both additive in contributions from carriers.

If a cylindrical band model is used, then

$$\frac{M_{xy}(0)}{H} = \frac{e^2 \tau_0^2 (T)}{\pi m^* c C_0 h^2} \; \frac{(kT)^{2p+2}}{T} \left\{ (2p+1) \; F_{2p} - (2p+2) F_{(2p+1)} \right\}. \quad (138)$$

This reduces in the case of extreme degeneracy and $p = 0$ to

$$\frac{M_{xy}(0)}{H} = \frac{\pi^2}{3\xi} \; \frac{n\mu^2 k}{c} \; , \quad (139)$$

so that

$$Q = \frac{\pi^2}{3\xi} \; \frac{k}{e} \; \frac{\mu}{c} \; . \quad (140)$$

Using a similar expression, Tamarin et al. [157] concluded that the measured coefficient was too large to be accounted for on the basis of diffusion terms alone, so that phonon-drag effects were important. However Spain and Platz concluded that Eq. (140) accounts for the majority of the experimentally measured coefficient [24] (Fig. 56).

In their measurements above about 80°K Mills et al. [161] found that the Nernst coefficient was independent of field to 10 kG. Theoretically calculated values [24] show that this is a general result when there are two bands with nearly equal mobilities contributing nearly equally to the tensor component M_{xy}. The field dependence observed by Tamarin et al. [157] below 80°K may then be attributed either to phonon-drag terms or to lack of compensation in their sample of graphite.

Shubnikov-de Haas oscillations can be observed in the Nernst coefficient at very low temperatures [96].

K. Electronic Contribution to Thermal Conductivity and Specific Heat

For most temperatures of interest the lattice thermal conductivity of graphite is many orders of magnitude greater than the electronic contribution. (For a review of the thermal conductivity see Ref. [126].) Klein, Straub, and Diefendorf [165] have shown that the electronic component is a substantial proportion of the total conductivity at low temperatures ($T \leq 10°K$). This component (K_e) can be considerably reduced by applying a magnetic field and can be related to the electrical conductivity through a generalized Wiedemann-Franz law [40, 166]:

$$K_e(H) = L_0 T \sigma(H),$$ (141)

where L_0 is the usual Lorenz number

$$L_0 = \frac{\pi^2}{3} \left(\frac{k}{e}\right)^2 = 2.45 \times 10^{-8} \ V^2/°K^2.$$ (142)

This result implies that the collisions that are effective in bringing the disturbed distribution function of the electron assembly back into momentum equilibrium (electrical conductivity) are equally effective in bringing it back into energy equilibrium (thermal conductivity). This condition is not always satisfied at low temperatures [124]. The result obtained by Holland et al. [166] may therefore be valid only in pyrolytic graphite, in which scattering at crystallite boundaries presumably predominates at low temperatures.

At very low temperatures ($T \leq 2°K$) the specific heat can be represented by an expression of the form [57].

$$C_v = \gamma T + aT^3,$$ (143)

where the T^3 term arises from the excitation of normal vibrational modes of the lattice and the linear term represents the electronic contribution to the specific heat. The parameter γ can be related to the density of states at the Fermi level [43]. Earlier measurements [167, 168] showed that the low-temperature behavior depended sensitively on the defect structure of the graphite and yielded values of γ ($\sim 24 \ \mu J/mole-°K^2$) that were much larger than the theoretical one ($13 \ \mu J/mole-°K^2$) [17]. It is probable that these measurements, made above $1°K$, were inaccurate because of the difficulty of extrapolating C/T to $0°K$. Later measurements to $0.4°K$ [169] yielded a truly straight-line plot for natural graphite from Ceylon, with a value of 13.8 $\mu J/\text{mole-}^\circ K^2$, in good agreement with theory.

VIII. C-AXIS TRANSPORT PROPERTIES

A. C-Axis Conductivity

There is still considerable controversy as to the intrinsic c-axis resistance. Measurements on synthetic materials gave a low conductivity value at $300^\circ K$ (~1-5 ohm^{-1} cm^{-1}), corresponding to an anisotropy ratio ($r = \sigma_{11}/\sigma_{33}$) greater than 10^3, whereas measurements on natural crystals gave values ranging from about 10^3 ($r = 100$) to 0.1 ohm^{-1} cm^{-1} ($r = 10^4$). Data are summarized in Table 18 and Figure 57.

Arguments have been put forward suggesting that the natural-crystal results of Primak and Fuchs [172] and Primak [173], in which the anisotropy ratio remains roughly constant down to $4^\circ K$ ($r = 100$), give the intrinsic ratio. There have been three major explanations put forward for the supposedly extrinsic effects then observed with higher anisotropy ratios. Young [175] has

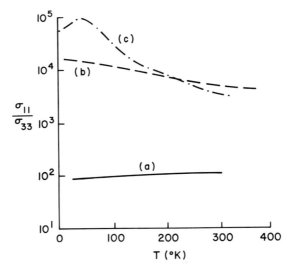

FIG. 57. Variation in the conductivity anisotropy ratio σ_{11}/σ_{33} as a function of temperature for synthetic and natural crystals: (a) natural crystal [172]; (b) annealed pyrolytic graphite [174]; (c) annealed, hot-pressed pyrolytic graphite [8].

TABLE 18

Comparison of Experimental Data on the c-Axis Resistivity of Graphite

Material	T ($^\circ$K)	ρ_{33} (ohm-cm)	(σ_{11}/σ_{33})	Remarks	Ref.
Natural crystals (Ceylon and Mexico)	300	1.2×10^{-3}	100		2
Natural crystals (Ceylon)	300	1.0	10^4		170
Natural crystals (Ceylon)	300	4 - 14	$10^4 - 10^5$	Anisotropy ratio studied as a function of purification procedure	171
Natural crystals (Ticonderoga)	300 77 4.2	$4-5.6 \times 10^{-3}$	110-170 150 120	σ_{11}/σ_{33} determined from potential distribution on surfaces as well as directly	172
Natural crystals (Ticonderoga)	300	5×10^{-3}	120	Very-high-quality crystals chosen	173
Pyrolytic graphite deposited at 2200°C	300	0.8	5,500	Hall effect, thermoelectric power, density, x-ray diffraction effects studied as a function of deposition temperature	148
Pyrolytic graphite deposited at 2500°C	300 4.2	1.2 1.8	5,000 3,500	Studies of Hall effect, magnetoresistance, X-ray diffraction as a function of deposition temperature	174
Pyrolytic graphite heat-treated to 3000°C	300 4.2	0.26 0.50	5,200 1.6×10^4		174
Hot-pressed pyrolytic graphite annealed at 3500°C	300 4.2	0.17 0.265	3,800 8.8×10^4	Basal galvanomagnetic characteristics very similar to Soule's natural crystals	8

TABLE 18-continued

Material	T ($^\circ$K)	ρ_{33} (ohm-cm)	$(\sigma_{11}/\sigma_{33})$	Remarks	Ref.
Natural crystals (Ticonderoga)	300	3×10^{-3}	80	Conductivity measured from time taken by electron to diffuse across the skin depth	107

effectively countered the "path-tortuosity" argument. Another model supposes that the material is made of slices of perfect material separated by thin amorphous regions which produce the enhanced resistance. The weaknesses of this model have been discussed by Spain [176]. A third model supposes that the stacking in synthetic graphite is either completely random or sufficiently disturbed to alter the band structure sufficiently to reduce curvature with respect to k_z. This disorder is not apparent from X-ray or electron-microscope studies except in grossly defective graphites, for which the anisotropy ratio is quite low (~200) [148, 174]. In addition, the band structure of synthetic material with a basal crystallite dimension greater than 1 μm is essentially the same as that of natural crystals [5, 8, 93].

It is in principle very easy to obtain a lower value for the experimentally determined anisotropy than a high value. In addition, it is possible that natural crystals contain non-basal dislocations [177] that can short out the high c-axis resistance along basal paths. Such defects are effectively removed in the conditions of plastic shear employed in the production of hot-pressed graphite [97]. Thus it is possible that the high values for the c-axis resistivity, confirmed by numerous workers on different materials, are a measure of intrinsic behavior.

In the event that the anisotropy ratio r is approximately 10^2, the mean free path for electron scattering is approximately 10^3 Å at 300°K, rising to approximately 10^4 Å at 4°K. This corresponds to mean c-axis mobilities at these temperatures of approximately 10^2 and 10^3 cm^2/V-sec, respectively. The relaxation time for scattering is then approximately isotropic. Clearly in this case the criterion Eq. (107) is satisfied, and the Boltzmann transport equation can be used to describe the conductivity. A calculation of the c-axis conductivity in this case has been made by Pospelov [178] and Haering and Wallace [179].

Klein [174] has also used this model to account for c-axis conductivity in synthetic material for which $r > 10^3$. In order to calculate c-axis mobilities, the number of free carriers is determined from basal galvanomagnetic effects and substituted into Eq. (120). Typical results are shown in Table 19, from which it can be seen that the calculated mean free path is approximately equal to the interlayer separation (2-5 Å), so that the mobility μ_{33} varies weakly with temperature. The anisotropy in the relaxation time is extremely high (τ_{11}/τ_{33} is ~30 at 300°K and ~700 at 4.2°K).

TABLE 19

Mobility, Relaxation Time, and Mean Free Path at Various
Temperatures for a Synthetic Sample[a]

Parameter	Value at		
	300°K	77.5°K	4.2°K
σ_{11} (ohm^{-1} cm^{-1}) x 10^4	2.26	3.87	33.2
σ_{33} (ohm^{-1} cm^{-1})	5.9	3.3	3.8
$(\sigma_{11}/\sigma_{33})$ x 10^4	0.38	1.2	8.8
μ_{11} (cm^2/V-sec) x 10^4	1.24	5.75	70
τ_{11} (sec) x 10^{-13}	3.5	16.2	196
ℓ_1 (Å) x 10^3	0.7	3.2	39
n_{total} (cm^{-3}) x 10^{18}	11.3	4.2	3.0
μ_{33} (cm^2/V-sec)	3.3	5.0	8.0
τ_{33} (sec) x 10^{-14}	0.95	1.6	2.7
ℓ_3 (Å)	0.95	1.6	2.7
τ_{11}/τ_{33}	32	100	730

[a]Data from Ref. [8], samples SA-26 and SC-15. For purposes of computation the following values were chosen: $m_{11}^* = 0.05 m_0$; $v_1 = 2$ x 10^7 cm/sec; $m_{33}^* = 6 m_0$; $v_3 = 10^6$ cm/sec [See Fig. 3].

It is instructive to analyze the experimental results at low temperatures. For the sample used in Table 19 the mean basal crystallite dimension is approximately 10^{-3} cm. Galvanomagnetic results indicate that scattering from the prism edges of these crystallites controls the relaxation time at 4.2°K, whereas Table 19 indicates that the carriers contributing to c-axis conductivity are scattered repeatedly as they cross the crystallites [176].

It is tempting to explain these results in terms of an anisotropic relaxation time for scattering. At low temperatures, impurity or carrier-carrier scattering may produce anisotropy of scattering in a single-electron ellipsoid, as in germanium and silicon [51]. At higher temperatures c-axis propagating modes with large wave-vector ($q_z = \pi/C_0$) may effectively control the momentum distribution disturbed by a field E_z but not appreciably affect the momentum distribution for basal current flow. The difficulties associated with this model are as follows:

1. The mean free path ($l_3 \approx 2\text{-}5$ Å) is too short for the Boltzman transport equation to be valid [Eq. (108)].

2. Relaxation-time anisotropies of the order cited have not been observed in other materials, even those with energy surfaces with principal axis ratios similar to those found in graphite (e.g., 12:1 for conduction electrons in germanium [51]).

3. The dependence of the mobility on temperature indicated in Table 19 is much too weak to be explained by the scattering mechanisms discussed.

The difficulties associated with the extreme anisotropy of the relaxation time may be resolved if an alternative model is adopted in which the carriers are restrained to free motion only along the basal planes [176]. With respect to such movement, the electron wavefunctions can be represented by delocalized Bloch-type wavefunctions, whereas in the third direction they are localized. This localization can be understood simply when the atomic-orbital configuration is considered. The σ bonds have a maximum charge-cloud density in the plane of the atoms, compared with the between-plane regions for the π bonds. If a small potential barrier $\Delta\epsilon$ separates the π electrons in different sheets, localization of the free carrier will occur with respect to the c-axis direction.

With this model, c-axis conductivity may occur when a carrier has sufficient energy to traverse the barrier. After collision with a phonon, a carrier may be accelerated across several interplanar spacings before further scattering reduces the energy sufficiently to produce localization within another sheet. This implies that the resistivity should approach an infinitely large value at low temperatures, whereas experimentally a maximum in the resistivity occurs at about 35°K. It is possible that at very low temperatures, where the anisotropy ratio is extremely high ($\sim 10^5$), basal shorting may effectively control the resistivity. This suggestion is strengthened by the observation that the c-axis pressure coefficient of resistivity is approximately equal to the basal coefficient at low temperatures but very different at high temperatures [139] (Fig. 37). Qualitatively the weak dependence on temperature of the c-axis resistivity can be understood at low temperatures from basal shorting effects and at higher temperatures from the balance between the increase of transition probabilities and the decrease of the mean free path as the phonon excitation is increased. At very high temperatures [$kT \gg \Delta\epsilon$ (T)] the ratio of the effective numbers of carriers taking part in c-axis conductivity and basal conductivity should approach unity, and the anisotropy ratio of conductivity (r) should approach a value compatible with normal conduction in all directions.

Young [175] has proposed that some degree of localization may be produced in grossly defective graphites from the destruction of long-range order, and in pyrolytic graphites above about 50°K from the interaction between electrons and c-axis propagating phonons. The model proposed by Spain [176] differs from this in that the localization is assumed to arise from electron-core and electron-electron interactions in the absence of structural defects or vibrational atomic displacements and therefore exists even at very low temperatures. It is possible that a model of the band structure of graphite, taking into account electron correlations in a meaningful way, will predict that the carriers are localized in one dimension. The Hubbard model [54] of the Hamiltonian in which electron correlations are included has already shown how localization in three dimensions may occur. It is also possible that electron-electron scattering in the limit of very small relaxation time and localization through electron-correlation terms are equivalent.

B. C-Axis Conductivity in Applied Magnetic Fields

The effect of applied magnetic fields on the c-axis conductivity has been studied at room temperature by Bhattacharya [180] and by Spain et al. [8].

between 4.2 and 300°K. One remarkable feature of the results is the magnitude of the longitudinal magnetoresistance $M_L^C = \Delta\rho_{33} (H_3)/\rho_{33} (0)$ compared with the transverse magnetoresistance $M_T^C = \Delta\rho_{33} (H_1)/\rho_{33} (0)$ (Figs. 58 and 60). The magnetoresistance ratio $M_L^C/M_T^C (=a)$ increases with specimen perfection at 300°K to values as high as 450. Since even the best quality synthetic

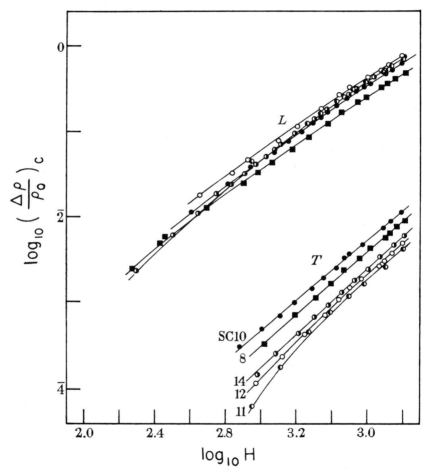

FIG. 58. Variation with magnetic field of the longitudinal and transverse magnetoresistance coefficients of graphite plotted on a log-log scale for several samples at 77°K. The transverse effect is much smaller than the longitudinal effect and does not share a common exponent in the magnetic field. From Ref. [8].

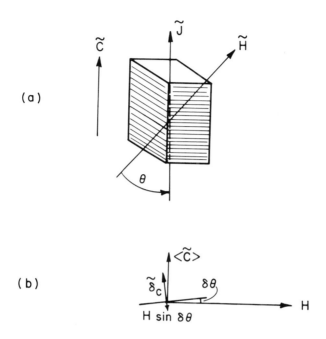

(a)

(b)

FIG. 59. (a) Geometrical arrangement of the magnetic field with respect to current flow for c-axis magnetoresistance effects. The longitudinal effect occurs for $\underset{\sim}{H} \mid \mid \underset{\sim}{c} \mid \mid \underset{\sim}{J}$, while the transverse effect occurs for $\underset{\sim}{H} \perp \underset{\sim}{c} \mid \mid \underset{\sim}{J}$. For both synthetic crystals (trigonal effects destroyed by random orientation of crystallites about the c-axis) and single crystals, the crystal symmetry is such that only the angle θ between the magnetic field vector $\underset{\sim}{H}$ and the c-axis $\underset{\sim}{c}$ needs to be specified. (b) Illustration of the effect of the "satellite" effects modifying the transverse c-axis magnetoresistance in a synthetic specimen. The vector $<\underset{\sim}{c}>$ represents the average c-axis vector direction for the whole crystal, with the magnetic field perpendicular to it. A crystallite with c-axis vector $\underset{\sim}{\delta_c}$ at an angle $\delta\theta$ to $<\underset{\sim}{c}>$ has a longitudinal effect M_C^L (H sin $\delta\theta$) that may be larger than the transverse effect M_C^T (H cos $\delta\theta$).

materials are characterized by small angular differences between the c-axis direction of the crystallites, the small transverse effects observed are partly due to satellite effects of the longitudinal magnetoresistance. This is so because the normal component of the magnetic field is never zero over the entire region of the crystal (Fig. 58).

However, the magnetoresistance ratio varies as a function of temperature (Fig. 60) indicating that this satellite effect does not dominate at low temperatures. Values for M_C^T at 4.2°K as high as 0.2 in 4 kG correspond to mobilities as high as 10^3 cm^2/V-sec if a normal model for compensated hole and electron conduction is assumed, in complete disagreement with the mean value given in Table 19. This large transverse coefficient could be interpreted in several ways. If the majority of the carriers are localized, it is possible that a small group of minority carriers contributes most of the c-axis conductivity, with the high mobility given above. Such a group of carriers would consist of less than 1% of the total number of carriers free to conduct along the basal planes at 4.2°K. Alternatively, if the conductivity at 4.2°K arises partly from basal shorting effects, as discussed in the preceding section, then the observed magnetoresistance may arise from the effect of the magnetic field on this current. It is emphasized that with the observed values of the c-axis conductivity the transverse magnetoresistance should be extremely small ($\lesssim 10^{-8}$ at 10 kG) if the mean carrier mobility is less than 1 cm^2/V-sec.

Similar difficulties are encountered in the interpretation of the longitudinal magnetoresistance. In a normal model of conduction processes this component is zero for spherical bands or for spheroidal bands with the magnetic field parallel to the principal axis [51]. Small effects can be explained when band warping is taken into consideration [51, 181]. Since the constant-energy surfaces in graphite are aligned along the vertical edges of the Brillouin zones, the measured coefficient should therefore be extremely small. The longitudinal magnetoresistance is clearly not a satellite effect of the basal transverse magnetoresistance, since the two effects do not share either a common temperature or magnetic field dependence. At very low temperatures, if basal shorting contributes to c-axis conductivity, the two effects are possibly mixed.

Recent measurements [149] of the longitudinal magnetoresistance M_L^C to fields of up to 80 kG indicate that saturation occurs (Fig. 61). Quantum oscillations are observed at temperatures as high as 77°K, with minima

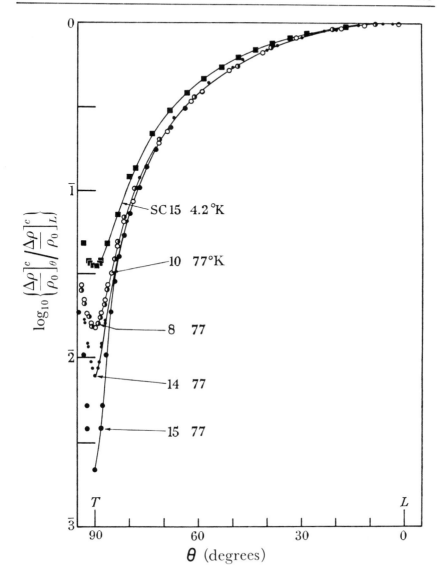

FIG. 60. The magnetoresistance ratio plotted logarithmically against θ, the angle between the magnetic field and the c-axis direction. The longitudinal to transverse magnetoresistance ratio increases with specimen perfection but changes with temperature; for example, for SC15, the ratio changes from 457 at 77°K to 25 at 4.2°K. From Ref. [8].

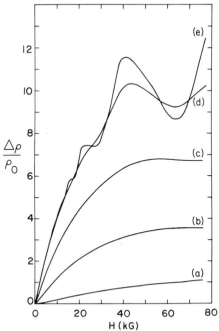

FIG. 61. Variation with magnetic field of the longitudinal c-axis
magnetoresistance coefficient up to 80 kG (8 teslas). The onset of Shubnikov-
de Haas oscillations and saturation can be seen clearly. The amplitude of the
oscillations compared with that of the classical effect (mid-line drift) is much
larger than that for the basal magnetoresistance for a sample cut from the
same material (recrystalized pyrolytic, Fig. 41). A small oscillation corresponding to the m = 1 electron level may be seen even at 77°K. (a) 305°K,
(b) 77°K, (c) 36°K, (d) 18°K, (e) 4.2°K. From Ref. [149].

agreeing with those observed in basal effects, but with much greater amplitude. Spin splitting of the m = 1 electron crossing can be seen clearly at
1.5°K, whereas the effect cannot be resolved in the basal conductivity of a
specimen cut from the same material (Figs. 41 and 62). The low-field oscillations have been used by Cooper and co-workers [22] to determine the effect
of neutron irradiation of the De Haas-Van Alphen periods $\Delta (1/H)$.

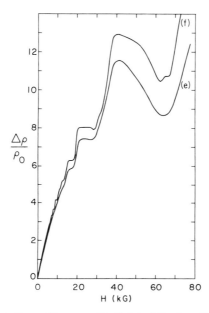

FIG. 62. Variation with magnetic field of the longitudinal c-axis magneto-resistance for the same sample as in Fig. 60 at (a) 4.2°K and (b) 1.5°K. The extra structure in the oscillations revealed at 1.5°K includes spin-splittings of the m = 1 electron and hole oscillations. Data taken from Ref. [149].

In summary, the c-axis conductivity and magnetoconductivity are not ᶠ⋯lly understood but afford an interesting field of study for both experimental ⋯ᵤd theoretical scientists.

IX. SUMMARY

In broad outline generalized band models account for the electronic properties of graphite obtained experimentally. The main thrust of recent work has been to fit experimental values to the overlap parameters appearing in the band models. Attempts are now being made to see whether a single set of band parameters can be found to fit all experimental information in detail. Since presently accepted band models are based on the one-electron approximation, it is not known whether a more general model including the effects of electron correlations is required to achieve this fit in detail.

There have been three developments of interest in the last few years:

1. The controversy as to the sign of γ_2 has been firmly resolved. The currently accepted negative sign will surely not require revision unless other ideas of the band structure change drastically.

2. The role that the parameter γ_4 plays in the band structure is clear, now that a definitive value (0.15-0.25 eV) has been assigned. This parameter appreciably affects the variation of carrier properties along the zone edge.

3. A definite value for the parameter γ_3 (~0.29 ± 0.02 eV) allows the extent of trigonal warping of the bands to be assessed realistically. It is unfortunate that such a large value of this parameter does not allow its effect to be calculated by using perturbation theory. Any realistic computation of the energy bands of graphite must now be made with the aid of a fast computer.

Our state of understanding of the electronic properties has largely arisen from advances in experimental techniques, enabling the Fermi surface and other characteristic properties of the electrons to be measured. Three experiments that may be singled out are cyclotron resonance, the de Haas–Van Alphen (and Shubnikov-de Haas) effect, and magnetoreflection.

It is likely that electron-correlation effects are required to explain completely the transport effects. Electron-electron scattering is presumably important in controlling the basal mobility. The mechanism of c-axis conductivity is still not satisfactorily resolved, although it is possible that the carriers are localized in basal sheets as a result of correlation effects. Since our knowledge of the electron bands in graphite and other solids comes largely from experiments at very low temperatures, it is not known what changes, if any, are required to describe the electronic properties at higher temperatures.

Despite initial interest, there is little possibility that graphite will be useful in thermoelectric devices. Current problems include the part played by phonon-drag effects in the thermoelectric power and other thermomagnetic coefficients. There is interest in the use of graphite as a model for plasma studies. It is anticipated that this area will see an increased experimental and theoretical activity in the next few years.

Finally, if interest in the electronic properties of graphite continues at its present level, it is interesting to speculate how much of the material in this chapter will be out of date in five years, or even two.

REFERENCES

[1]   D. E. Roberts, Ann. Phys., 40, 453 (1913).

[2]   G. E. Washburn, Ann. Phys., 48, 236 (1915).

[3]   C. A. Coulson, Nature, 159, 265 (1947).

[4]   P. R. Wallace, Phys. Rev., 71, 622 (1947).

[5]   S. J. Williamson, S. Foner, and M. S. Dresselhaus, Phys. Rev.,
      140A, 1429 (1965).

[6]   A. R. Ubbelohde and F. A. Lewis, Graphite and Its Crystal Compounds
      Oxford Univ. Press, London and New York, 1960.

[7]   I. L. Spain, A. R. Ubbelhode, and D. A. Young, in Proc. 2nd Ind.
      Carbon and Graphite Conf. Soc. Chem. Ind., p. 123.

[8]   I. L. Spain, A. R. Ubbelohde, and D. A. Young, Phil. Trans. Roy.
      Soc. (London), A262, 1128 (1967).

[9]   J. Linderberg, Arkiv Fysik, 30, 557 (1965).

[10] J. Linderberg and K. V. Mäkila, Solid State Commun., 5, 353 (1967).

[11] R. R. Haering and S. Mrozowski, Prog. Semi-Conductors, V, 273
      (1960).

[12] J. C. Slonczewski and P. R. Weiss, Phys. Rev., 109, 272 (1958).

[13] W. N. Reynolds, Physical Properties of Graphite, Elsevier,
      Amsterdam, 1968.

[14] Le Groupe Français D'Étude des Carbones, Les Carbones, Vol. 1,
      Masson, Paris, 1965.

[15] C. A. Klein, in Chemistry and Physics of Carbon (P. L. Walker, Jr,
      ed.), Vol. 2, Dekker, New York, 1966, p. 225.

[16] S. Ergun, in Chemistry and Physics of Carbon (P. L. Walker, Jr.,
      ed.), Vol. 3, Dekker, New York, 1968, p. 45.

[17] J. W. McClure, IBM J. Res. Develop., 8, 255 (1964).

[18] J. W. McClure, in Proc. Intern. Conf. Phys. Semimetals and
      Narrow-Band Semiconductors, Dallas, 1970, Pergamon, London,
      1971, p. 127.

[19] G. A. Saunders, in Modern Aspects of Graphite Technology (L. C. F.
      Blackman, ed.), Academic Press, New York, 1970, p. 79.

[20] P. R. Schroeder, Dresselhaus, M. S. and A. Javan, Phys. Rev.
      Letters, 20, 1292 (1968).

[21] J. A. Woollam, Phys. Letters, 32A, 115 (1970).

[22] J. D. Cooper, J. P. Smith, J. Woore, and D. A. Young, J. Phys.,
      C4, 442 (1971).

[23] I. L. Spain, J. Chem. Phys., 52, 2763 (1970).

[24] I. L. Spain and C. Platz, Bull. Amer. Phys. Soc., 15, 574 (1970)
      and work to be published.

[25] V. V. Kechin, Fiz. Tverd. Tela, 11, 1788 (1970) [Sov. Phys.--Solid
      State, 11, 1448 (1970)].

[26] R. F. Willis, B. Feuerbacher, and B. Fitton, Phys. Rev., B4,
      2441 (1971).

[27] R. B. Aust and H. G. Drickamer, Science, 140, 817 (1963).

[28] H. G. Drickamer, Science, 156, 1183 (1967).

[29] J. D. Bernal, Proc. Roy. Soc. (London), A160, 749 (1924).

[30] R. R. Haering, Can. J. Phys., 36, 352 (1958).

[31] J. W. McClure, Carbon, 7, 425 (1969).

[32] R. W. G. Wyckoff, Crystal Structures, Vol. 1 Interscience, New
      York, 1960.

[33] C. A. Coulson and H. Taylor, Proc. Phys. Soc. (London), A65,
      815 (1952).

[34] D. F. Johnston, Proc. Roy. Soc. (London), A227, 349 (1955).

[35] D. F. Johnston, Proc. Roy. Soc. (London), A237, 48 (1956).

[36] W. M. Lomer, Proc. Roy. Soc. (London), A227, 300 (1955).

[37] F. J. Corbato, in Proc. 3rd Conf. Carbon, Pergamon, New York
      1957, p. 173.

[38] W. Van Haeringen and H. G. Junginger, Solid State Commun., 7, 1723
      (1969).

[39] C. S. Painter and D. E. Ellis, Phys. Rev., 1B, 4747 (1970).

[40] J. M. Ziman, Electrons and Phonons, Oxford University Press, 1960.

[41] J. Kolodziejcak, L. Soznowski, and W. Zawadski, Rept. Intern.
      Conf. Semi-Cond. Phys., Exeter, 1962.

[42] S. Ono and K. Sugihara, J. Phys. Soc. Japan, 24, 818 (1968).

[43] J. W. McClure, Phys. Rev., 108, 612 (1957).

[44] M. S. Dresselhaus and J. G. Mavroides, Carbon, 3, 465 (1966).

[45] P. R. Schroeder, M. S. Dresselhaus, and A. Javan, in Proc. Intern. Conf. Phys. Semimetals and Narrow-Band-Gap Semiconductors, Dallas, 1970, Pergamon, London, 1971, p. 139.

[46] G. Dresselhaus and M. S. Dresselhaus, Phys. Rev., 140, A401 (1965).

[47] J. W. McClure and Y. Yafet, in Proc. 5th Conf. Carbon, Vol. 1, Pergamon, New York, 1962, p. 12.

[48] J. W. McClure, unpublished results, 1971.

[49] D. F. Johnston, in Appendix to W. H. Reynolds and P. R. Goggin, Phil. Mag., 5, 1049 (1960).

[50] C. A. Klein, J. Appl. Phys., 35, 2947 (1964).

[51] A. C. Beer, in Solid State Physics, Suppl, 4 (F. Seitz and D. Turnbull eds.), Academic Press, New York, 1963.

[52] A. H. Wilson, The Theory of Metals, 2nd ed., Cambridge University Press, 1953.

[53] D. Pines, Elementary Excitations in Solids, Benjamin, New York, 1964.

[54] J. Hubbard, Proc. Roy. Soc. (London), A296, 82, 100 (1967).

[55] S. Raimes, The Wave Mechanics of Electrons in Metals, North-Holland, Amsterdam, Chapter 10.

[56] P. B. Visscher and L. M. Falicov, Phys. Rev., B3, 2541 (1971).

[57] C. Kittel, Introduction to Solid State Physics, Wiley, New York, 1966, Chapter 10.

[58] J. M. Ziman, Principles of the Theory of Solids, Cambridge University Press, 1964.

[59] A. D. Boardman and D. G. Graham, J. Phys., C2, 2320 (1969).

[60] C. Kittel, Quantum Theory of Solids, Wiley, New York, 1963.

[61] W. A. Harrison, Solid State Theory, McGraw-Hill, New York, 1970.

[62] F. Bassani and G. P. Parravicini, Nuovo Cimento, 50, 95 (1967).

[63] E. Doni and G. P. Parravicini, Nuovo Cimento, 64B, 117 (1969).

[64] A. B. Pippard, The Dynamics of Conduction Electrons, Gordon and Breach, London, 1965.

[65] M. S. Dresselhaus and J. G. Mavroides, IBM J. Res. Develop., 8, 262 (1964).

[66] J. W. McClure, Phys. Rev., 119, 606 (1960).

[67] M. Inoue, J. Phys. Soc. Japan, 17, 808 (1962).

[68] Y. Uemara and J. Inoue, J. Phys. Soc. Japan, 13, 382 (1958).

[69] J. M. Luttinger and W. Kohn, Phys. Rev., 97, 869 (1965).

[70] J. W. McClure, Phys. Rev., 104, 666 (1956).

[71] P. Nozieres, Phys. Rev., 109, 1510 (1958).

[72] P. Nozieres, in Proc. 3rd Conf. Carbon, Pergamon, New York, 1959, p. 197.

[73] J. McDougal and E. Stoner, Phil. Trans. Roy. Soc. (London), 237, 67 (1938).

[74] A. C. Beer, M. N. Chase, and P. F. Choquard, Helv. Phys. Acta., 28, 529 (1955).

[75] P. Rhodes, Proc. Roy. Soc. (London), A204, 396 (1950).

[76] R. G. Arkhipov, V. V. Kechin, A. I. Likhter, and Yu. A. Pospelov, Zh, Eksp. Teor. Fiz., 44, 1964 (1963) [Sov. Phys.--JETP, 17, 1321 (1963)].

[77] S. Ono and K. Sugihara, J. Phys. Soc. Japan, 21, 861 (1966).

[78] K. Sugihara and S. Ono, J. Phys. Soc. Japan, 21, 631 (1966).

[79] J. A. Woollam, Phys. Rev. Letters, 25, 810 (1970).

[80] J. K. Galt, W. A. Yager, H.W. Dail, Phys. Rev., 103, 1586 (1956).

[81] B. Lax and H. Zieger, Phys. Rev., 105, 1466 (1957).

[82] S. J. Williamson, M. Surma, H. C. Praddaude, R. A. Patten, and J. K. Furdyna, Solid State Commun., 4, 37 (1966).

[83] G. F. Dresselhaus, M. S. Dresselhaus, and J. G. Mavroides, Carbon, 4, 433 (1966).

[84] D. E. Soule, J. W. McClure, L. B. Smith, Phys. Rev., 134, 453 (1964).

[85] D. Schoenberg, Phil. Trans. Roy. Soc. (London), 245, 1 (1952).

[86] T. Berlincourt and M. Steele, Phys. Rev., 98, 956 (1955).

[87] D. E. Soule, Phys. Rev., 112, 698 (1958).

[88] D. E. Soule, Phys. Rev., 112, 708 (1958).

[89] W. J. Spry and P. M. Scherer, Phys. Rev., 120, 826 (1960).

[90] D. E. Soule, IBM J. Res. Develop., 8, 268 (1964).

[91] E. S. Itskevitch and L. M. Fisher, Zh. Eksp. Teor. Fiz. Pis'ma, 5, 141 (1967) [Sov. Phys.--JETP Letters 5, 114 (1967)].

[92] J. R. Anderson, W. J. O'Sullivan, J. E. Schirber, and D. E. Soule, Phys. Rev., 164, 1038 (1967).

[93] J. A. Woollam, Phys. Rev., B4, 3393 (1971).

[94] S. J. Williamson, S. Foner, and M. S. Dresselhaus, Carbon, 4, 29 (1966).

[95] R. B. Dingle, Proc. Roy. Soc. (London), A211, 517 (1952).

[96] J. A. Woollam, Phys. Rev., B3, 1148 (1971).

[97] A. W. Moore, A. R. Ubbelohde, and D. A. Young, Proc. Roy. Soc. (London), A280, 153 (1964).

[98] D. E. Soule, Proc. 5th Conf. Carbon, Pergamon, New York, 1961, p. 13.

[99] E. N. Adams and T. D. Holstein, J. Phys. Chem. Solids, 10, 254 (1959)

[100] V. V. Kechin, A. I. Likhter, and G. N. Stepanov, Fiz. Tverd. Tela, 10, 1242 (1968) [Sov. Phys.-Solid State, 10, 987].

[101] D. J. Flood, Phys. Letters, 30A, 168 (1969).

[102] I. L. Spain, Proc. Conf. Electronic Density of States, 1969, N.B.S. Publication, Washington, D.C., p. 717.

[103] K. S. Krishnan and N. Ganguli, Nature, 144, 667 (1939).

[104] J. W. McClure, J. Chim. Phys. (France), 57, 859 (1960).

[105] J. W. McClure and L. B. Smith, Proc. 5th Conf. Carbon, Vol. 2, Pergamon, New York, 1961, p. 3.

[106] P. N. Argyres, Phys. Rev., 117, 315 (1960).

[107] G. Wagoner, Phys. Rev., 118, 3, 647 (1960).

[108] J. A. Woollam, Phys. Letters, 32A, 371 (1970).

[109] A. C. Baynham and A. D. Boardman, Plasma Effects in Semiconductors--Helicon and Alfvén Waves, Taylor and Francis, London, 1971.

[110] M. Surma, J. K. Furdyna, and H. C. Pradduade, Phys. Rev. Letters, 13, 710 (110).

[111] S. Kakihana and L. Buchmiller, J. Appl. Phys., 38, 5376 (1967).

[112] W. Boyle and P. Nozieres, Phys. Rev., 111, 782 (1958).

[113] S. Yashinsky and S. Ergun, Carbon, 2, 355 (1965).

[114] Y. Sato, J. Phys. Soc. Japan, 24, 489 (1968).

[115] L. B. Leder and J. A. Suddeth, J. Appl. Phys., 31, 1422 (1960).

[116] K. Zeppenfeld, Phys. Letters, 25A, 335 (1967).

[117] K. Zeppenfeld, Z. Phys. ik, 211, 391 (1968).

[118] Y. H. Ichikawa, Phys. Rev., 109, 653 (1958).

[119] E. A. Taft and H. R. Phillip, Phys. Rev., 138, A197 (1965).

[120] D. L. Greenaway, G. Harbeke, F. Bassani, and E. Tosatti,
       Phys. Rev., 178, 1340 (1969).

[121] S. Ergun, J. B. Yashinsky, and J. R. Townsend, Carbon, 5, 403 (1967).

[122] J. G. Carter, R. H. Huebner, R. N. Hamm, and R. D. Birkhoff,
       Phys. Rev., 137, A639 (1965).

[123] J. W. McClure, J. Chim. Phys., 57, 859 (1960).

[124] F. J. Blatt, Solid State Phys., 4, 199 (1961).

[125] J. M. Radcliffe, Proc. Phys. Soc. (London), A68, 675 (1955).

[126] B. T. Kelly, in Chemistry and Physics of Carbon (P. L. Walker, Jr.,
       ed.), Vol. 5, Dekker, New York, 1969, p. 119.

[127] Yu. A. Pospelov and V. V. Kechin, Fiz, Tverd. Tela, 5, 3574
       (1964) [Sov. Phys.--Solid State, 5, 2622].

[128] K. Sugihara and H. Sato, J. Phys. Soc. Japan, 18, 332 (1963).

[129] K. Komatsu, J. Phys. Soc. Japan, 10, 346 (1955).

[130] A. Yoshimori and Y. Kitano, J. Phys. Soc. Japan, 10, 346 (1956).

[131] J. G. Collins, Phys. Rev., 155, 663, (1967).

[132] H. Frohlich, Proc. Roy. Soc. (London), A188, 532 (1947).

[133] H. Frohlich and B. V. Paranjabe, Proc. Phys. Soc. (London), B69,
       21 (1956).

[134] C. Herring, reported in P. Debye and E. Conwell, Phys. Rev., 93,
       693 (1954).

[135] R. Keyes, J. Phys. Chem. Solids, 6, 1 (1958).

[136] T. P. McLean and E. G. S. Paige, J. Phys. Chem. Solids, 16, 220
       (1960).

[137] T. P. McLean and E. G. S. Paige, J. Phys. Chem. Solids, 18, 139
       (1961).

[138] K. Sugihara, J. Phys. Soc. Japan, 21, 324 (1966).

[139] M. Yeoman and D. A. Young, J. Phys., C2, 1742 (1969).

[140] J. W. McClure and W. J. Spry, Phys. Rev., 165, 809 (1968).

[141] R. A. Morant, J. Phys. (London), D3, 1367 (1970).

[142] J. W. McClure, Phys. Rev., 112, 715 (1958).

[143] D. E. Soule and J. W. McClure, J. Phys. Chem. Solids, 8, 29 (1959).

[144] P. W. Bridgman, Proc. Amer. Acad. Arts Sci., 81, 165 (1947).

[145] G. A. Samara and H. G. Drickamer, J. Chem. Phys., 37, 471 (1962).

[146] A. I. Likhter and V. V. Kechin, Fiz. Tverd. Tela, 5, 3066 (1964) [Sov. Phys.--Solid State, 5, 2246.].

[147] G. H. Kinchin, Proc. Roy. Soc. (London), A217, 9 (1953).

[148] L. C. F. Blackman, G. A. Saunders, and A. R. Ubbelohde, Proc. Roy. Soc. (London), A264, 19 (1961).

[149] I. L. Spain and J. A. Woollam, Sol. St. Comm., 9 1581 (1971) and further work to be published.

[150] H. J. Goldsmid and J. M. Corsan, Phys. Letters, 8, 221 (1964).

[151] S. Mizushima and T. Endo, J. Phys. Soc. Japan, 24, 1402 (1968).

[152] S. Yugo, H. Hayashi, and K. Yazawa, Appl. Phys. Letters, 17, 339 (1970).

[153] Y. Baskin and L. Meyer, Phys. Rev., 100, 544 (153).

[154] L. F. Vereshchagin and S. S. Kabalkina, Dokl. Akad, Nauk SSSR, 131, 300 (1960) [Sov. Phys. - DOKL., 5, 373].

[155] J. D. Cooper, J. Woore, and D. A. Young, Nature, 225, 721 (1970).

[156] T. Takezawa, T. Tsuzuku, A. Ona, and Y. Hishiyama, Phil. Mag., 19, 623.

[157] P. V. Tamarin, S. S. Shalyt, and V. I. Volga, Fiz, Tverd, Tela, 11, 1725 (1969) [Sov. Phys.--Solid State, 11, 1399].

[158] K. Sugihara, J. Phys. Soc. Japan, 29, 1465 (1970).

[159] J. P. Jay-Gerin and R. Maynard, J. Low Temp. Phys., 3, 377 (1970).

[160] K. Sugihara, T. Takezawa, T. Tsuzuku, Y. Hishiyama, and A. Ono, 12th Intern. Conf. Low Temp. Phys., Kyoto, 1970.

[161] J. J. Mills, R. A. Morant, and D. A. Wright, Brit. J. Appl. Phys., 16, 479 (1965).

[162] D. A. Wright, Brit. J. Appl. Phys., 14, 329 (1963).

[163] H. J. Goldsmid and D. E. Lacklison, Brit. J. Appl. Phys., 16, 573 (1965).

[164] I. M. Tsidilkovskii, Thermomagnetic Effects in Semiconductors, Infosearch, 1960.

[165] C. A. Klein, W. D. Straub, and R. J. Diefendorf, Phys. Rev., 125, 468 (1962).

[166] M. Holland, C. Klein, and W. Straub, J. Phys. Chem. Solids, 27, 903 (1966).

[167] P. H. Keesom and V. Pearlman, Phys. Rev., 99, 1119 (1955).

[168] W. DeSorbo and G. E. Nichols, J. Phys. Chem. Solids, 6, 352 (1958).

[169] B. J. C. Van der Hoeven and P. H. Keesom, Phys. Rev., 130, 1318 (1963).

[170]   A. K. Dutta, Phys. Rev., 90, 187 (1953).

[171]   R. Bhattacharya, Indian J. Phys., 33, 407 (1959).

[172]   W. Primak and L. H. Fuchs, Phys. Rev., 95, 22 (1954).

[173]   W. Primak, Phys. Rev., 103, 544 (1956).

[174]   C. A. Klein, J. Appl. Phys., 33, 3338 (1962).

[175]   D. A. Young, Carbon, 6, 135 (1968).

[176]   I. L. Spain, in Proc. Intern. Conf. Phys. Semimetals and Narrow-
        Band-Gap Semiconductors, Dallas, 1970 (D. L. Carter and R. T.
        Bate, eds.), Pergamon, London, 1971, p. 177.

[177]   C. Roscoe and J. D. Thomas, Proc. Roy. Soc. (London), A297, 397
        (1967).

[178]   Yu. A. Pospelov, Fiz, Tverd. Tela, 6, 1525 (1964) [Sov. Phys.--
        Solid State 6, 1193].

[179]   R. R. Haering and P. R. Wallace, J. Phys. Chem. Solids, 3, 253
        (1957).

[180]   R. Bhattacharya, Indian J. Phys., 34, 53 (1965).

[181]   A. B. Pippard, Proc. Roy. Soc. (London), A282, 464 (1964).

[182]   D. B. Fischbach, Phys. Rev. 123, 1613 (1961).

[183]   G. Hennig, B. Smaller, and E. Yasaitis, Phys. Rev., 95, 1088
        (1954).

[184]   J. K. Galt, W. A. Yager, and F. R. Merritt, Proc. 3rd Conf.
        Carbon, Pergamon, New York, 1959, p. 193.

[185]   A. Balzarotti and M. Grandolfo, Phys. Rev. Letters, 20, 9 (1968).

[186]   K. K. Kobayashi and T. Uemura, J. Phys. Soc. Japan, 25, 404
        (1968).

ACKNOWLEDGMENTS

The author wishes to thank Prof. P. L. Walker Jr. for encouragement in writing this review, and Prof. A. R. Ubbelohde, F.R.S. and Dr. D.A. Young for giving him his initial interest in the subject. Discussions during the writing of the manuscript are gratefully acknowledged with Profs. J. R. Anderson, M. S. Dresselhaus, J. W. Mc Clure and Dr. J. A. Woollam. Thanks are also due to Mr. R. O. Dillon for correcting manuscripts and suggesting improvements. A grant from the U. S. Atomic Energy Commission is gratefully acknowledged.

SURFACE PROPERTIES OF CARBON FIBERS

D. W. McKee and V. J. Mimeault

General Electric Research and Development Center
Schenectady, New York

I. INTRODUCTION

Interest in carbon fibers and filaments extends back for many years. In his classic work on the incandescent lamp Edison made carbon filaments before 1880 by carbonizing natural cellulose fibers, such as cotton and linen. A further development occurred in 1909, when Whitney patented a process for coating carbon fibers from cellulose with pyrolytic graphite by flashing at temperatures up to 4000°C. However, after the introduction of tungsten filaments, interest in carbon for lamp applications declined.

In the 1950s the search for new materials for structural composites generated an upsurge of interest in carbon fibers. Early work at this time on pyrolyzed viscose rayon at Wright-Patterson Air Force Base produced relatively strong flexible fibers by stretching the fiber during carbonization at 2000°C; however, the process was not very reproducible. Low-strength carbon and graphite yarns and fabrics were produced, mainly by the Union Carbide Corp., in the early 1960s and found application as tape windings for rocket nose cones and heat shields.

A significant breakthrough in carbon-fiber technology occurred in the period 1960-1965, when it was discovered independently by R. Bacon, A. Shindo, and W. Watt that very-high-strength carbon filaments could be obtained by subjecting the precursor fiber to a continuous tensile stress during the high-temperature treatment. It is the high values of the specific modulus and specific strength, due in part to low density (~2.0 g/ml), that make these new carbon fibers attractive materials as structural reinforcing agents.

At present there are two main types of commercially available high-strength carbon fiber. These are derived from a-cellulose [e.g., Thornel (Union Carbide Corp.) and Hitco (Thompson Co., Inc.)] and polyacrylonitrile (PAN) [RAE fiber, Fortafil (Great Lakes Carbon Co.), Modmor (or Morganite, Morganite Ltd.), and Grafil (Courtaulds Ltd.)], respectively. Experimental quantities of high-strength fiber have also been produced from Saran (a copolymer of vinyl chloride and vinylidene chloride), polyvinyl alcohol, polyimide, and even from pitches and asphalt. The manufacturing process of the commercial fiber varies in detail depending on the precursor polymer and the heat-treatment cycle used. Typically the process involves the following basic steps:

1. Preoxidation at a low temperature (200-300°C) to promote cross-linking, thereby stabilizing the precursor.

2. Carbonization in an inert atmosphere at about 1000°C.

3. Graphitization at 2500 to 3000°C.

The rayon-based fibers are usually stressed at graphitization temperatures, whereas the polyacrylonitrile-based fibers are stressed during the low-temperature preoxidation step. Both precursors lose 50 to 80% of their initial weight during high-temperature heat treatment. In some cases (e.g., Modmor type II) the final heat treatment is carried out at a lower temperature (~1200°C), resulting in a fiber of lower modulus but higher tensile strength. As discussed in Section III, the stretching process orients the graphitic fibrils or crystallites with their basal planes parallel to the fiber axis, the degree of orientation being of major significance in determining the mechanical properties of the fiber.

A distinction is often made between "carbon" and "graphite" fibers. Thus Bacon [1] defines "carbon" fibers as those obtained by heat treatment to a maximum temperature of 1000 to 1500°C. These often contain residual precursor material, whereas "graphite" fibers have been heated to above 2500°C and are at least 99% carbon. Ruland and associates [2] have pointed out that as the structure of "graphite" fibers is not identical in a crystallographic sense with natural graphite, the use of the term "graphitization" should not be interpreted in a structural sense, but rather as referring only to fibers that have been heat-treated to temperatures exceeding 2000°C. In this review only the term "carbon fibers" will be used to avoid confusion.

The outstanding mechanical properties of carbon fibers become of practical interest only if they can be efficiently translated into a usable structural form, such as a composite. Although the tensile properties of carbon fibers appear to be directly related to the size and orientation of the graphitic subunits, the properties of a fiber-resin composite depend to a large extent on the adhesion between the fiber and the matrix. Composite properties highly dependent on fiber-resin coupling include flexural and

interlaminar shear strengths and modes of failure. Early work with carbon
fiber-epoxy composites gave disappointingly low values for the interlaminar
shear strength, a general measure of the adhesion at the interface. As a
result considerable effort was subsequently devoted to changing the surface
condition of the fibers and thereby increasing the strength of the bond between fiber and resin matrix.

The surface properties of the carbon fibers, the changes that can be
induced in the fiber surface and their effect on composite properties, and
comparison with surface properties of other forms of carbon are the subjects
of this chapter. Only investigations relevant to this specialized topic have
been included. The discussion is limited to the high-strength fibers produced during the last decade, and, although both "carbon" and "graphite"
types are included, no attempt has been made to review earlier work on
"soft" carbon filaments or yarns.

As much useful information on the properties of high-strength carbon
fibers still remains inaccessible in the form of proprietary or classified
reports, this review is in no sense complete. The aim has been rather
to present a balanced picture of the nature and properties of the carbon-fiber
surface and the significance of these properties in determining the behavior
of the resulting composites. The rapidly expanding technical literature in
this area has been reviewed up to the end of 1969.

II. EXPERIMENTAL METHODS FOR THE STUDY OF
CARBON-FIBER SURFACES

A large number of techniques have been used to investigate the properties of carbon fibers. The more important methods that can give direct
information concerning the fiber surface are discussed in this section.

A. X-Ray Diffraction

X-Ray diffraction techniques have been widely used to probe the structure of carbon fibers and especially to obtain detailed information on important morphological variables, such as the degree of graphitization and
crystallite size and orientation.

Several excellent reviews of X-ray studies of carbon and graphite have
recently appeared [3, 4], and the basic theory will not be repeated here.
Typical applications to fiber technology include the identification of

graphitic structure in Modmor by Brydges and associates [5], who used the
position of the (002) reflection to obtain the interlayer spacing. The presence
of mixed X-ray reflections also suggested that the layers were packed in a
turbostratic arrangement.

A detailed X-ray study of PAN-based fibers was recently carried out
by D. J. Johnson and Tyson [6], who used both high-angle and low-angle
methods. A typical high-angle diffraction photograph of the carbon fibers
showed the (002) reflection oriented on the equator and the (100) partially
oriented on the meridian. Lack of three-dimensional order was indicated by
the absence of other hkl reflections with l > 0, thus confirming the presence
of turbostratic stacking.

The position and integral width of the (002) reflection curve gave the
layer spacing c and the crystallite dimension L_c after correction for various
effects, such as incoherent scattering, strain broadening, and polarization.
The value of L_c can also be obtained from the half-maximum peak height
width, which gives $1.13L_c$. The (100) reflection can be analyzed by the
methods developed by Franklin and Warren [7] to give values for the crystallite dimensions L_c and L_a corresponding to the c and a directions, respectively.

The degree of preferred orientation of the crystallites in carbon fibers
can also be determined from the intensity of reflection along the (002) diffraction arc. The angular half-width of the resulting bell-shaped intensity
distribution, measured at the half-maximum intensity, gives a measure of
possible misorientation about the fiber axis. Thus Bacon orientation functions $I(\phi)$ [8], showing the distribution of crystallite c-axis orientations with
respect to the fiber axis, can be obtained [5, 9] (Fig. 1).

For the purpose of estimating the preferred orientation in carbon fibers
Yamamoto and Yamada [10] define a coefficient of orientation, α_{10}, by the
relation

$$\alpha_{10} = \frac{\int_0^{10°} \pi \, F \, (\phi)^2 \, d\phi}{\int_0^{90°} \pi \, F \, (\phi)^2 \, d\phi}, \tag{1}$$

where

$$F(\phi) = \frac{I(\phi)}{I(0)} = \frac{I_{obs}(\phi)}{I_{obs}(0)} , \qquad\qquad (2)$$

$$I_{obs}(\phi) = I(\phi)f^2 LA, \qquad\qquad (3)$$

where $I(\phi)$ is the true intensity, $I_{obs}(\phi)$ is the observed intensity, f is the structure factor, L is the Lorentz factor, and A is the absorption factor. By plotting the angle ϕ against the ratio of intensities $F(\phi)$, the orientation coefficient a_{10} could be calculated.

An X-ray diffractometer adapted for use with carbon fibers has been described by Logsdail [9]. This arrangement, which uses a toroidal mirror, is shown schematically in Fig. 2.

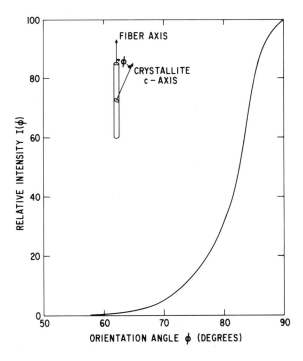

FIG. 1. X-Ray intensity profile for Modmor type I fibers. After Brydges et al. [5].

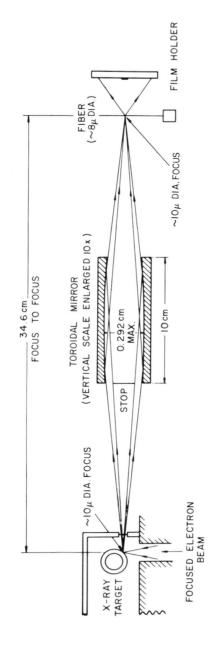

FIG. 2. X-Ray diffraction arrangement for study of carbon fibers. After Logsdail [9].

The theory of line broadening and preferential orientation in fibers has been developed recently by Ruland [11]. Low-angle scattering has been considerably used to obtain information on the sizes of crystallites and the shapes and sizes of voids in carbon fibers. These methods are derived mainly from the treatments developed by Debye and Bueche [12] and Porod [13].

D. J. Johnson and Tyson [6] investigated the porosity and internal surface of PAN-based fibers by low-angle scattering. According to Debye and Bueche [12], the intensity of the scattered radiation is given by

$$I(ks) = 4\pi \int_0^\infty r^2 \, C(r) \, \frac{\sin ksr}{ksr} \, dr, \tag{4}$$

where $ks = 4\pi(\sin \theta)/\lambda$ and $C(r)$ is the correlation function that gives the probability that a line of length r in the fiber will have both ends situated in pores. If $I^{-1/2}$ is taken proportional to θ^2, which is often found to be the case for carbon fibers, then

$$C(r) = \exp\left(\frac{-r}{a}\right), \tag{5}$$

where a is called the correlation length. Hence

$$\left(\frac{I_{\theta=0}}{I}\right)^{1/2} = 1 + \frac{16\pi^2\theta^2 a^2}{\lambda^2} \tag{6}$$

and a can be determined from the slope and intercept of the plot of $I^{-1/2}$ against θ^2. The term a is related to the specific surface area S by

$$S = \frac{4c(1-c)}{a}, \tag{7}$$

where c is the packing density or fraction of the total volume occupied by carbon. Since c can be estimated from the density ratio of the carbon fiber and solid graphite, the surface area S can be evaluated. As this value is computed from measurements on powdered fibers, the result may be very different from that obtained by gas adsorption on whole or chopped fibers (see Section IV).

The mean chord intercept length $\bar{l}_p$ for the voids or pores within the carbon fibers is defined [6] by

$$\bar{l}_p = \frac{4(1-c)}{S} = \frac{a}{c} \tag{8}$$

and hence can also be evaluated from the plot of $I^{-1/2}$ against θ^2.

According to the treatment developed by Porod [13], the surface area is also given by the relation

$$S = \frac{2\pi^2 c(1-c)}{\lambda} \quad \frac{\lim\limits_{\infty} [(2\theta)^4 \, I(2\theta)]}{\int_0^\infty (2\theta)^2 \, I(2\theta) \, d(2\theta)} \tag{9}$$

and S can be determined from this integral by a numerical method [6].

Information concerning the shapes of micropores in carbon fibers can be obtained from the curves of $I\theta^4$ against 2θ and $I\theta^3$ against 2θ. In a further development of Porod's theory Tchoubar-Vallat and Méring [14] have shown that a distinction can be made between sharp-edged (e.g., parallelepipedal) and smooth-edged (e.g., cylindrical) pores, the former giving a limiting value in the plot of $I\theta^3$ against 2θ, and the latter producing oscillations about the limiting value. This method has been used by D. J. Johnson and Tyson [6] to demonstrate that polyacrylonitrile-based fibers contain sharp-edged voids.

In a further development of low-angle X-ray scattering the intensity distribution has been related to the distribution of intercrystalline separation normal to the fiber axis [6].

Other applications of X-ray diffraction to carbon fibers include studies of the development of graphitization and the relation between the structure of the carbonized fiber and that of the precursor [15, 16], and studies of the epitaxial deposition of crystalline monomers on rayon-based fibers [17].

B. Electron Diffraction

Standard electron-diffraction techniques have been widely used to study crystallite orientation and three-dimensional order within carbon fibers [18] and to demonstrate the relationship between modulus and preferred orientation [19]. For studying the surface layers of the fiber, electron diffraction is generally a more suitable tool than X-ray diffraction, as the electrons only penetrate to a depth of about 1000 Å into the surface, whereas X rays penetrate the entire fiber.

Electron-diffraction experiments by W. Johnson and Watt [20] and D. J. Johnson and Tyson [6] demonstrated that the (002) layer planes of the crystallites lie closely parallel to the fiber axis for high-strength PAN-based fibers, the general arrangement of the layers being turbostratic. A typical electron-diffraction pattern for a PAN-based fiber is shown in Fig. 3,

FIG. 3. Electron-diffraction pattern for Courtaulds B fiber.

the (002) and diffuse (100) reflections being clearly evident. By ion-etching removal of the original surface layers, it has been shown by electron diffraction that the graphitic layer structure is not a surface phenomenon.

Other applications of electron diffraction include the use of this technique to study the dependence of the degree of graphitization on heat treatment and on degree of stretch during fabrication of rayon-based fibers [18]. The recrystallization of nickel-graphite fibers on heat treatment has also been studied by this method [6]. Partially crystalline material intermediate in structure between turbostratic and three-dimensional graphite was detected.

C. Electron Microscopy

Transmission electron microscopy has been used by many investigators to study the fine structure of carbon fibers. For example, D.J. Johnson and Tyson [6] sectioned PAN-based fibers longitudinally with a Huxley microtome and observed the presence of fine fibrils and needlelike pores. A typical electron micrograph of a PAN-based fiber in longitudinal section is shown in Fig. 4. Information on possible subgrain boundaries can also be obtained from observations of the extinction bands that appear orthogonal to the crystallite direction [6].

FIG. 4. Electron micrograph (x58,000) of Thornel 40 fiber in longitudinal section Courtesy of R.B. Bolon and associates [21].

Dark-field micrographs using the (002) reflections of the diffraction
pattern have been used by Badami et al. [22] to estimate crystallite size and
by Butler and Diefendorf [23] to show that the surface crystallites are often
larger in size than the internal ones. A phase-contrast method developed
by Heidenreich et al. [24] has recently been used by Fourdeux et al. [18] to
obtain images of the stacks of graphitic crystallites in the fibrils of rayon-
based fibers.

D. Scanning Electron Microscopy

The increased depth of focus and contrast of the scanning electron microscope are being widely used to reveal structural and surface features of carbon fibers and composites. Recent applications of this instrument include studies of the structure of fiber-resin composites [25], the fracture surfaces of composites [26], the surface and internal flaws and voids in single fibers and their role in fracture [27], and the interface between metal and fiber in metal-fiber composites [28].

E. Spectroscopy

An interesting method of determining the orientation of the crystallites in carbon fibers by electron spin resonance (ESR) has recently been reported by Robson et al. [29]. Bundles of PAN-based fibers, cut to 2 to 3 mm in length, were carefully packed into the microwave cavity of a conventional X-band ESR spectrometer with 100-Hz phase-sensitive detection. As the static magnetic field was rotated about an axis perpendicular to the fiber axis, a systematic variation in the g value of the single broad resonance line was observed. As shown in Fig. 5, the g value was a maximum when the magnetic field was aligned perpendicular to the fiber axis, and a minimum when the field was parallel to this axis. The g values in fact showed a variation of the form

$$g = g_\perp + A \cos^2 \theta, \tag{10}$$

similar to that found for graphite single crystals [30], shown in Fig. 6. It can therefore be concluded that the crystallites in the fiber were preferentially aligned with their basal planes parallel to the fiber axis. Although no distribution of c-axis orientations can be obtained by this method, the ESR results generally confirm the conclusions of X-ray and electron diffraction with respect to preferred orientation.

High-energy photoelectron spectroscopy has recently been applied to the study of surface structures in carbon fibers by Barber et al. [31]. The method involves measurement of the binding energies of electrons in the 1s levels of carbon, oxygen, and nitrogen in the surface layers. Shifts in the energies of the core electrons are sensitive to the chemical environment of the atoms, and information concerning the functional groups present on the fiber surface can in principle be obtained. Typical results for a Grafil

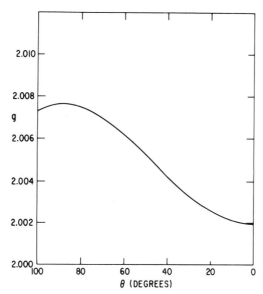

FIG. 5. g-Value anisotropy of carbon-fiber electron spin resonance.
After Robson et al. [29].

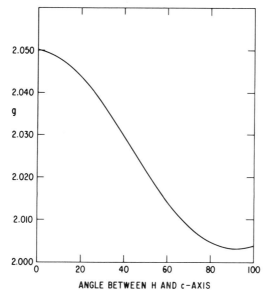

FIG. 6. g-Value anisotropy of crystalline-graphite electron spin
resonance. After Wagoner [30].

type II fiber that had been baked to 1000°C in nitrogen are shown in Fig. 7
(top spectrum). The spectrum for the same fiber after an oxidation treatment is also shown (bottom spectrum). Comparison of the spectra reveals a
marked increase in surface-bound oxygen after the oxidation treatment, the
oxygen being held in at least two different forms. Appreciable amounts of
nitrogen atoms are also present on the fiber surface, although these could
be almost completely removed by heating to 2000°C.

Trace amounts of surface impurities in carbon fibers can also be detected by Auger electron spectroscopy, which is useful for identifying the
elements present in the first few atomic layers of the fiber surface. Results
obtained by Connell [32] for a Modmor II fiber sample are shown in Fig. 8,
in which the derivative of the energy distribution dN(E)/dE of the emitted
electrons is plotted against the electron energy. Trace amounts of phosphorus,
sulfur, chlorine, and oxygen are detectable on the fiber surface.

Information on the crystal structure of carbon-fiber surfaces has
recently been obtained by Raman spectroscopy, which can be used to distinguish between different types of carbon and graphite. Raman spectra for
Modmor I and II, obtained by Tuinstra and Koenig [33], are shown in Fig. 9.
The observed differences in line width indicate that Modmor II has essentially a carbon surface, whereas Modmor I is more graphitic, with larger
surface crystallites than those in the type II fiber.

F. Contact Angle and Wetting

1. Direct Measurement of Contact Angle

Owing to the difficulties involved in handling small filaments, few
contact-angle measurements have been made with carbon fibers.

Bobka and Lowell [34] have described a method that has been used for
fibers of glass and carbon. The pressure drop Δp across a curved liquid-
vapor interface is given by the Young-Laplace equation:

$$\Delta p = \gamma \left(\frac{1}{R_1} + \frac{1}{R_2} \right), \tag{11}$$

where R_1 and R_2 are the radii of curvature of the interface and γ is the surface tension. For surfaces of revolution (Fig. 10) the equation of the
meniscus takes the form

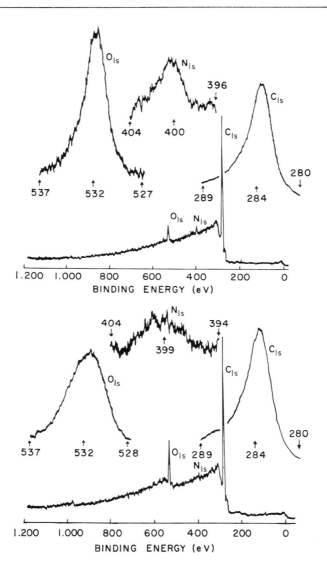

FIG. 7. High-energy-photoelectron spectra of 1s nitrogen, oxygen, and carbon electrons in Grafil II carbon fiber. Top spectrum, fiber baked at 1000°C in nitrogen. Bottom spectrum, fiber baked in nitrogen and then oxidized. After Barber et al. [31].

$$\frac{d^2z/dr^2}{[1 + (dz/dr)^2]^{3/2}} + \frac{dz/dr}{r[1 + (dz/dr)^2]^{1/2}} - \frac{z}{a^2} = 0, \tag{12}$$

where z is the rise of liquid at distance r from the axis of the filament. The term a is given by

$$a = \frac{\gamma}{g \, \Delta\rho} \, , \tag{13}$$

g being the acceleration of gravity and $\Delta\rho$ the difference in density between the liquid and vapor.

A photomicrograph of the interfacial region between fiber and liquid is made, the optical axis of the microscope being normal to the fiber axis and in the plane of the liquid surface. A fourth-order Runge-Kutta approximation can be used to give the best fit of the equation to the observed (r, z) values of the liquid profile. The contact angle θ can be computed by finding the slope of the liquid at the fiber surface.

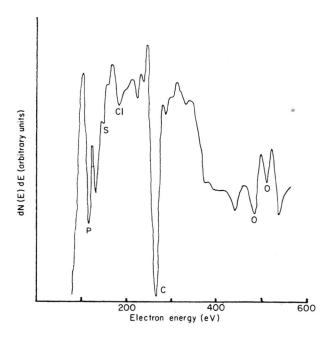

FIG. 8. Auger electron spectra of Modmor II fiber. After Connell [32].

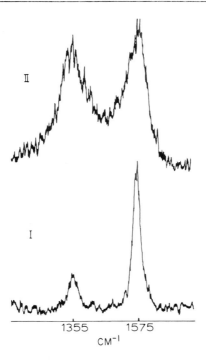

FIG. 9.   Raman spectra of Modmor I and II fibers.   After Tuinstra and Koenig [33].

2.   Wicking Rates

More general information on wetting characteristics of fibers can be obtained by measuring wicking rates.   Two methods have been used by Bobka and Lowell [34] in investigations on graphitic yarn.

a.   Mass-Pickup Method.   A yarn sample is suspended from the arm of an analytical balance, and the lower end of the sample is brought into contact with the surface of a suitable liquid by means of an adjustable jack. After a fixed time interval (e.g., 15 sec) the dish containing the liquid is lowered, and the change in the weight of the sample is recorded.   This procedure is repeated several times until the weight of the sample becomes constant.   This rather rough method provides general information concerning the extent of fiber-liquid interaction.

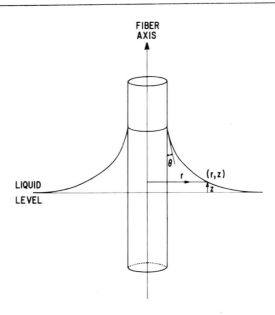

FIG. 10.   Profile of liquid surface near carbon fiber.   After Bobka and Lowell [34].

b. Surface-Velocity Method. A sample of graphite yarn is placed between the metal prongs of the cell shown in Fig. 11. A single thickness of
absorbent paper is inserted between the yarn and one of the prongs. The
change in electrical resistance across the prongs when the tissue becomes
wetted with liquid is measured by an external circuit.

Liquid is initially added to the reservoir arm of the cell, which is immersed in a constant-temperature bath. The liquid is brought to a desired
level by introducing gas under pressure through one of the reservoir inlets.
A pair of leads at S, connected to a recorder, are arranged to record a
signal when the stopcock is closed to stop the rise in liquid level. This
signal serves to mark the start of the wicking time interval. When liquid
initially wets the paper, a deflection is observed on the recorder. The
distance between the yarn-paper interface and the liquid level is measured
with a cathetometer to give the surface velocity of the liquid film.

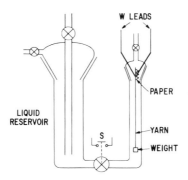

FIG. 11. Apparatus for determining wicking rates. After Bobka and Lowell [34].

For a system of uniform capillaries Washburn [35] derived an equation for the rate of flow dh/dt in a capillary of radius r:

$$\frac{dh}{dt} = \frac{r\gamma \cos\theta}{4\eta h} - \frac{r^2 g\rho}{8\eta} , \qquad (14)$$

where ρ, γ, and η are the liquid density, surface tension, and viscosity, respectively, θ is the contact angle, and g is the acceleration due to gravity.

The treatment of Peek and McLean [36] is a better approximation for a yarn bundle:

$$\frac{dh}{dt} = \frac{[2\xi \int_{-\infty}^{\infty} r^{-1} \phi(r)\, dr] - h\rho g}{8\eta h \int_{-\infty}^{\infty} r^2 \phi(r)\, dr \int r^{-4} \phi(r)\, dr} , \qquad (15)$$

where $\phi(r)\, dr$ is the fraction of the total number of channels with a radius between r and r + dr, and the parameter $\xi = \gamma \cos\theta$. The equation may be written in the form

$$\frac{dh}{dt} = \frac{A\xi}{4\eta h} - \frac{B\rho g}{8\eta} , \qquad (16)$$

where A and B are constants with values that depend on the function $\phi(r)$. If all the channels have radius r and $\xi = \gamma \cos\theta$, this equation reverts to Washburn's equation, Eq. (14).

By neglecting the influence of gravity on the wicking rate, one can obtain a simplified equation of the form

$$\frac{dh}{dt} = \frac{K}{h} \tag{17}$$

or, on integration, $h^2 = 2Kt$ for the initial condition that $h = 0$ when $t = 0$. This equation was found to fit the observed wicking-rate data quite well, particularly at small liquid heights, when the contribution of gravity could be reasonably neglected.

G. Surface-Area Measurements

Although an estimate of internal surface area can be obtained from the results of low-angle X-ray-scattering studies [6], most surface-area measurements have been made by the conventional BET method with nitrogen as adsorbate at $-195^{\circ}C$ [37, 38].

Because of the low (generally less than 2 m^2/g) surface area of carbon fibers, the use of krypton as an adsorbate at $-195^{\circ}C$ is often more convenient and accurate. Mimeault and McKee [39] used a low-dead-space apparatus similar to that described by Rosenberg [40], the pressure of the adsorbate being measured by means of a calibrated thermistor gauge.

Some surface areas have been made by a flow method utilizing the Perkin-Elmer Sorptometer (Model 212C) [41, 42]. Although this method is not capable of high accuracy for surface areas below 5 m^2/g, it has been used to measure gross changes in surface area resulting from oxidative surface treatment. In this technique nitrogen is adsorbed from a flowing mixture of helium and nitrogen, the sample being cooled to $-195^{\circ}C$. The volume of nitrogen subsequently desorbed on warming the sample to room temperature in a stream of pure helium is then recorded by means of a thermal conductivity cell. By repeating the measurements at various nitrogen partial pressures, a BET plot can be constructed and hence the surface area computed.

H. Adsorption Measurements

Apart from surface-area measurements, few adsorption isotherms have been determined on carbon fibers. Isotherms of organic vapors (mainly aromatic and aliphatic hydrocarbons) have been measured on Saran carbon fibers [43] by a gravimetric method, similar to that used by Everett and Whitton [44], which involves recording the weight changes of the sample to within 0.1 mg by means of a quartz helical spring from which the sample is suspended.

I. Porosity

Estimates of pore size and shape in carbon fibers can be obtained from X-ray-scattering data [6, 45] and electron microscopy [6] (see Sections II. A and II. B).

Didchenko [37] and others [46] have attempted to measure pore-size distributions in the range 0.02 to 100 μm by means of an Aminco-Winslow mercury porosimeter (Model 5-7109). However, because of the small average pore size of the fibers, little useful information can be obtained by this method. Meaningful mercury densities can, however, be measured by this technique [46].

Nitrogen pore-size distributions in the 0-600-$\overset{\circ}{A}$ range have been measured [37] from nitrogen adsorption isotherms at -195°C, using the Barrett-Joyner-Halenda [47] method of computation, as modified by Roberts [48]. It is still doubtful, however, that the smallest pores are accessible to the nitrogen molecule.

III. SURFACE STRUCTURE OF CARBON FIBERS

A. Surface and Internal Structure

Since the time of their first discovery the structure of high-strength carbon fibers has received much attention. Early work by Tang and Bacon [15] revealed the progressive development of oriented graphitic layers in cellulose fibers during pyrolysis. Figure 12 shows X-ray diffraction patterns of a regenerated-cellulose fiber (Fortisan-36, Celanese Corp.) after carbonization at successively higher temperatures. Pattern B shows that the crystal structure of the fiber remains essentially intact after 12-h heat treatment at 245°C. Above this temperature, however, the cellulosic structure begins to break down, becoming amorphous at 305°C (pattern D). At 900°C a graphitic structure begins to appear, and at 2800°C a preferred orientation of the graphite layers along the axis of the fibers is evident (pattern F).

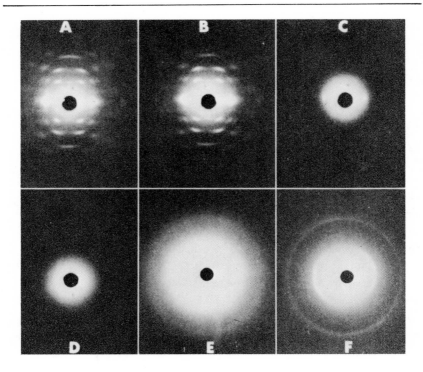

FIG. 12. X-Ray diffraction patterns of a regenerated-cellulose fiber
(Fortisan-36) before (A) and after (B-F) carbonization at increasing temperatures. Patterns B, C, D, E, and F were obtained after heat treatment
at 245, 280, 305, 900, and 2800°C, respectively. Reprinted by permission
from Ref. [15].

Tang and Bacon [15] also derived a possible mechanism for the condensation of cellulose rings into graphite layers. Thus, as shown schematically in
Fig. 13, the residue from each cellulose ring was believed to be a four-
carbon unit that served as a basic building block for the development of
graphite layers. There thus appeared to be a direct correspondence between
the molecular arrangement in the cellulose precursor and the preferred
orientation of the carbonized fiber. The graphitic layers in these low-
strength fibers were thus thought to be long and narrow, with the basal planes
oriented nearly parallel to the fiber axis and enclosing a system of elongated
pores.

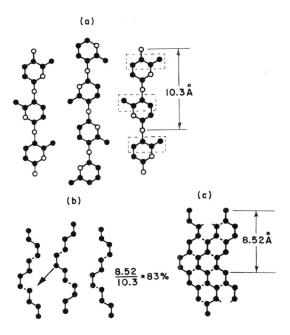

FIG. 13. Possible scheme for condensation of cellulose-ring units into graphitic layers. Reprinted by permission from Ref. [15].

The surface topography of early rayon-based graphitized fibers has been discussed by Badami and co-workers [16]. After pyrolysis to above $2500°C$, the fibers had very smooth surfaces, in contrast with carbonized fibers formed at lower temperatures. The internal structure appeared to be granular, with grains of about 40 Å in size. Although these fibers were not completely graphitized, low rates of heating were found to promote preferential orientation of the graphitic layers.

The announcement of the discovery of high-strength carbon fibers by W. Johnson and Watt [20] was accompanied by X-ray- and electron-diffraction information concerning their structure. These PAN-based fibers appeared to possess a fibrillar network structure, the fibril units being about 100 Å wide and aligned roughly parallel to the fiber axis. X-Ray diffraction patterns showed that the graphitic crystallites were turbostratic in structure, the dimension L_c being at least 12 layer planes, and L_a in the range

60 to 120 Å. As with the low-strength cellulose-based fibers studied by
Tang and Bacon [15], it appeared that the general structure of the graphitized
fiber could be related to that of the PAN parent fiber.

Essentially the same conclusions were reached by Badami, Joiner, and
Jones [22] from structural studies of another PAN-based fiber (Modmor I).
In this case the graphite crystallite size was estimated to be 50 Å from X-
ray line broadening, the basal planes being oriented along the fiber axis,
with a spread of $\pm 10^{\circ}$. The d spacing of the graphite crystals in these fibers
was 3.39 Å, compared with 3.35 Å for natural graphite, although no three-
dimensional reflections were observed. In this case also electron micrographs showed evidence of fibrils running parallel to the fiber axis.

Although most authors have interpreted the structure of high-strength
PAN-based fibers in terms of a fibrillar network, a somewhat different
model has recently been proposed by D.J. Johnson and Tyson [6] for these
materials. The idealized structure suggested by these authors is illustrated
schematically in Fig. 14. Arrays of misoriented turbostratic graphitic
crystallites were believed to be stacked end to end to give a columnar

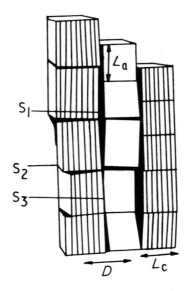

FIG. 14. Model of PAN-based carbon-fiber structure, according to
D.J. Johnson and Tyson [6]. Symbols: S_1, void; S_2, subgrain twist boundary;
S_3, intercrystallite boundary. Reprinted by permission from Ref. [6].

arrangement with sharp-edged voids separating the stacks. The average
width of the crystallites was estimated to be 65 Å. The frequency distribution of width, measured from the lengths of extinction bands, is shown in
Fig. 15. The mean width of the pores was less than 10 Å. In addition to
the turbostratic graphite, a perfectly stacked three-dimensional graphite
phase was also found. This component was especially common in fibers
that had been recrystallized in the presence of nickel and iron [49-51]. As
the turbostratic crystallites shown in Fig. 14 are separated by tilt and twist
boundaries running transverse to the fiber axis, this model is fundamentally
distinct from the fibrous network proposed earlier.

In connection with the structure of PAN-based fibers it should be
pointed out that the possession of a well-oriented graphitized structure is
confined to those fibers that have been heat-treated at temperatures in excess of 2500°C. Fibers prepared at lower temperatures may have more
complex and less highly ordered structures. For example, Modmor II fiber,
prepared by heat treatment of polyacrylonitrile between 1000 and 2000°C,

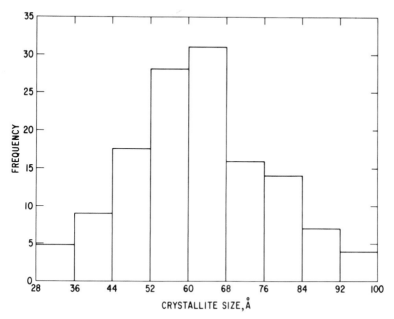

FIG. 15. Frequency distribution of crystallite width in PAN-based
carbon fibers. After D.J. Johnson and Tyson [6].

shows only poorly developed crystallinity. A phase-contrast transmission electron micrograph of a sample of Modmor II (Fig. 16) indicates that graphitization has proceeded only as far as the formation of isolated nuclei in which the arrangement of the layers is still somewhat irregular. By

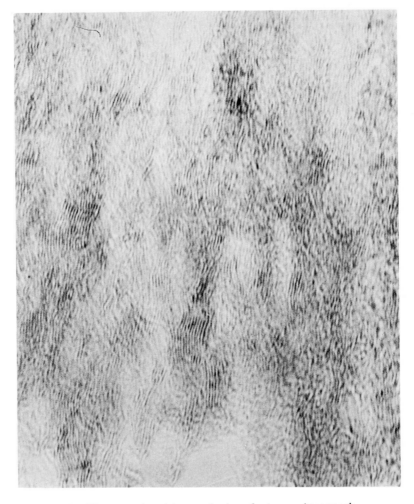

FIG. 16.   Phase-contrast transmission electron micrograph (x ~ 2,300,000) of Modmor II fiber.   Courtesy of R. B. Bolon et al. [21].

contrast, Modmor I, prepared at higher temperatures, has the much more regular structure shown in Fig. 17, the graphitic layer being aligned parallel to the fiber axis over long distances. A smooth transition from the structure shown in Fig. 16 to that of Fig. 17 has been observed by annealing Modmor II fibers at progressively higher temperatures [21].

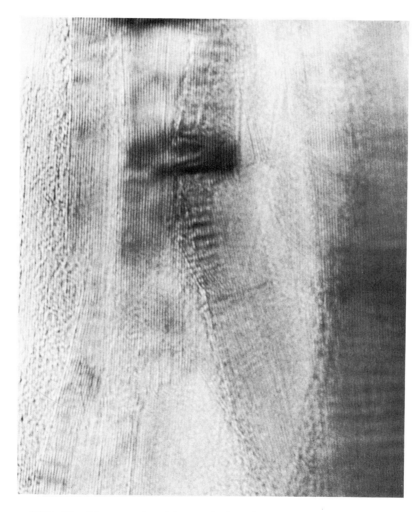

FIG. 17. Phase-contrast transmission electron micrograph (x ~2,300,000) of Modmor I. Courtesy of R.B. Bolon et al. [21].

Detailed studies of graphitized rayon (cellulose) fibers by means of small-angle X-ray scattering have recently been made by Ruland and co-workers [18, 52, 53]. Of special significance in this work are the conclusions regarding the preferred orientation of graphitic crystallites in the fibers as a function of heat treatment and modulus. In general the results again favor a fibrillar model, the main axis of the fibrils lying parallel to the fiber axis. The diameters of the fibrils range from 50 to 100 Å, and the lengths in excess of 1000 Å [18]. This structure gives rise to a network of needlelike pores, 10 to 20 Å in diameter and more than 200 Å in length, aligned along the fiber axis [53]. The porosity of these fibers is markedly dependent on the temperature of heat treatment, the average diameter of the voids increasing from 6 to 20 Å as the heat-treatment temperature is raised from 900 to 2900°C. Stretching, however, has no effect on the mean pore size, although high degrees of stretch cause ultimate collapse of the pores. The preferential orientation of the graphite crystallites with basal planes parallel to the fiber axis is also confirmed by the X-ray studies of Ergun [54]. Electron spin resonance has also been used to show that growth and ordering of the crystallites increase with heat-treatment temperature [29].

The graphitic surface structure of rayon-based carbon fibers has recently been observed directly by Hugo, Phillips, and Roberts [55], who used a Thornel 50 fiber with a fluted cross section and found it possible to pass an electron beam through regions at the periphery of the fiber to give electron micrographs such as that shown in Fig. 18. The arrangement of basal planes parallel to the fiber surface is clearly indicated here, the measured interlayer spacing being 3.4 Å. A striking feature of this structure is the continuity of the graphite basal planes over long distances, in spite of sharp bends in individual packets of planes. No evidence of normal or subgrain boundaries was observed in this work, the graphitic layers resembling continuous ribbons. A schematic representation of this model of the fiber structure is shown in Fig. 19. The granular layer at the surface of the fiber shown in Fig. 18 appears to be polyvinyl alcohol sizing applied to improve handling properties.

Finally, there is some evidence that some carbon fibers have a composite structure. Thus Butler and Diefendorf [23], using dark-field electron microscopy, have shown that the surface crystallites of PAN-based fibers are larger than the internal ones. The alignment of the crystallites with

FIG. 18. Transmission electron micrograph of Thornel 50 fiber. After Hugo et al. [55].

respect to the fiber axis also appears to be more nearly perfect in the outermost layers.

In summary, although there remain subtle differences in the proposed internal structure of high-strength carbon fibers, there seems little doubt that the surfaces of fibers produced at 2500°C and higher temparatures are

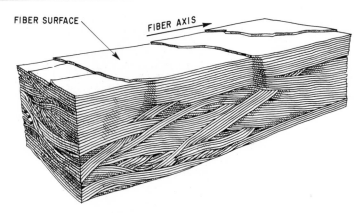

FIG. 19. The Roberts [21] model of Thornel fiber structure. Courtesy
of B.W. Roberts.

essentially graphitic in nature and almost pure basal plane in the case of
the strongest fibers. Fibers produced at lower temperatures are less
clearly characterized, and their surfaces may contain larger regions of
amorphous carbon or randomly oriented graphitic layers. The graphitic
nature of the surface of high-strength carbon fibers suggests that studies
of the surface energy, wetting, topography, reactivity, and functional
groups of graphite-crystal surfaces are relevant in interpreting the properties of carbon-fiber surfaces. These topics will be discussed briefly in the
sections that follow.

B. Topography of Carbon Fibers

The physical structure of carbon-fiber surfaces is of considerable importance in determining the interfacial area available for bonding between
fiber and matrix in a composite structure. Optical and electron micrographs
reveal that carbon fibers prepared by different manufacturers often show
quite distinct topographical features. A scanning electron micrograph of a
fracture surface of a Thornel-epoxy composite is shown in Fig. 20. An
irregular fluted cross-section is shown by this type of fiber, the surfaces
of which are characterized by longitudinal grooves and striations.

By contrast PAN-based fibers usually show a circular cross section
with a relatively smooth and featureless topography, like that in Fig. 21,

FIG. 20.   Scanning electron micrograph of the fracture surface of a
Thornel-epoxy composite.   Courtesy of R. B. Bolon et al. [21].

which shows the end and surface of a Modmor I fiber.   The circular cross
section of another type of PAN-based fiber (Courtauld B) is shown in Fig. 22.

The effect of surface treatments on the surface morphology and resulting
composite properties will be discussed in Section V.

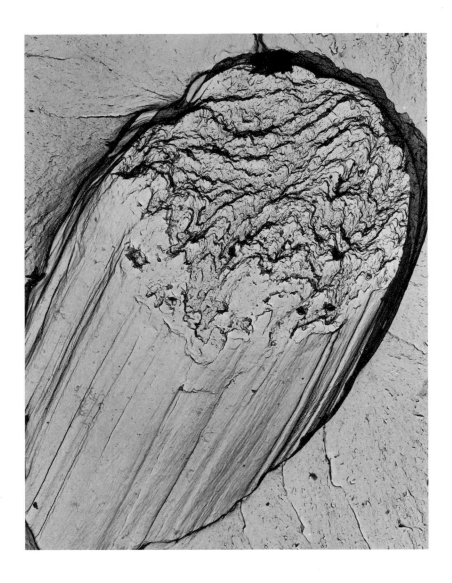

FIG. 21. Electron micrograph (x9600) of a replicated fracture surface of Modmor I fiber. Courtesy of R. B. Bolon et al. [21].

FIG. 22. Scanning electron micrograph of the fracture surface of a
Courtaulds B-epoxy composite. Courtesy of R.B. Bolon et al. [21].

IV. SURFACE AREA AND POROSITY OF CARBON FIBERS

The surface areas of carbon fibers are routinely measured by the BET
method with nitrogen or krypton as adsorbates. In particular, changes in
surface area are widely used to monitor the effects of various surface
treatments, as described in Section V. On the other hand, information on

porosity is often limited to measurements of total pore volume and indirect evidence from low-angle X-ray scattering (Section III). The available data are summarized here for the most common types of high-strength fibers.

A. Cellulose-Based Fibers

Early work on carbon and graphitized fibers obtained from a-cellulose (34, 46] showed differences of two orders of magnitude or more between the surface areas of carbon fibers and those of fibers that had been graphitized at high temperatures. This difference is illustrated in Table 1 [34].

TABLE 1

Properties of Fibers from a-Cellulose[a]

Property	Carbon fiber	Graphitized fiber
Surface area (m^2/g)	260	3
Open porosity (ml/g)	0.22	0.017

[a]From Ref. [34].

Electron micrographs of these fiber surfaces also indicated that carbon fibers formed at low temperatures are highly porous, with rough surfaces, the effect of graphitizing being to reduce the porosity open to the surface and anneal out much of the surface roughness. Considerable increase in the surface area of the graphitic yarn could, however, be produced by etching with acid sodium dichromate solution, the surface area increasing from 3 to 23 m^2/g after 24-h treatment with this etchant.

The effect of heat treatment on the properties of carbon yarn was investigated by Pallozzi [56]. The original material (Union Carbide's VYB 105-115) had a surface area of 340 m^2/g and contained appreciable amounts of hydrogen and oxygen surface complexes. As shown in Table 2, on successive heat treatments the carbon content increased, whereas the hydrogen and oxygen contents decreased, as did the surface area. Although the tensile strengths of these fibers were not appreciably affected, loss in surface area and decrease in surface functional groups were accompanied by a decrease in shear strength of composites made from these fibers, as shown in Fig. 23.

TABLE 2

Effect of Heat Treatment on the Properties of Carbon Yarn[a]

Property	Sample			
	1	2	3	4
Carbon content (%)	86.5	89.3	96.0	99.2
Hydrogen content (%)	0.65	0.31	0.31	0.5
Oxygen content (%):				
As CO	2.08			0.10
As CO_2	1.35			0.01
Surface area (m^2/g)	340	185	75	1-2

[a]Data from Ref. [56]. Carbon yarn was VYB 105-115, Union Carbide Corp.

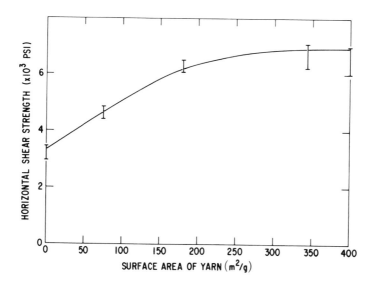

FIG. 23. Effect of fiber surface area on the shear strength of a composite mode from epoxy and carbon yarn (VYB 105-115). After Pallozzi [56].

More detailed studies of surface area and pore-size distribution in the Thornel series of high-strength fibers have been carried out recently [37]. The reproducibility of these data on Thornel 40 is shown in Table 3, and a typical pore-size distribution histogram, obtained from nitrogen-desorption isotherms, is shown in Fig. 24.

TABLE 3

Precision of Surface-Area and Porosity Measurements on Thornel 40[a]

Run	Surface area (m^2/g)	Pore volume (ml/g)	Pore diameter ($\overset{\circ}{A}$)		
			Upper quartile	Median	Lower quartile
1	1.3	0.0033	550	170	62
2	1.4	0.0030	375	135	57
3	1.6	0.0031	375	135	53
4	1.2	0.0029	475	135	57
Mean	1.38	0.00306	444	144	57
Standard deviation	0.17	0.00017	85	18	3.6

[a]From Ref. [37].

The basic differences in porosity and surface area between high-strength Thornel fibers and low-strength carbon yarns are illustrated in Table 4, where the properties of the two Thornels are compared with those of a low-temperature carbon fiber. The latter shows a much larger total pore volume and surface area than the high-strength fibers, which, however, exhibit a larger mean pore diameter than the yarn, the pores of which seem to be less than 20 $\overset{\circ}{A}$ in diameter. It is interesting that the calculated geometrical surface area of both Thornel fibers is about 0.57 m^2/g, that is, about one-third of the total area. In the case of the carbon yarn the large surface area can only be accounted for on the basis of a well-developed microporosity.

Surface-area values of about 1 m^2/g for Thornel fibers have also been reported by Goan and Prosen [57] and Scola and Brooks [58]. This can

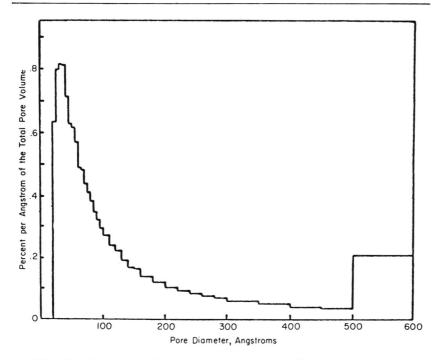

FIG. 24. Pore-size distribution for Thornel 40 fiber. From Ref. [37].

TABLE 4

Surface Area and Porosity of Thornel Fibers[a]

Material	Pore volume (ml/g)	Surface area (m²/g)	Pore diameter 0-600 Å			Mercury density (g/ml)
			Upper quartile	Median	Lower quartile	
Thornel 25	0.0033	1.6	475	155	63	1.46
Thornel 40	0.0030	1.3	473	166	67	1.59
Carbon yarn	0.129	281	All less than 20 Å			1.52

[a]From Ref. [37].

apparently be increased by an order of magnitude or more by suitable oxidative treatments [41] (see Section V).

In addition to the macropores discussed here, small-angle X-ray studies
of rayon-based fibers [53] reveal the presence of needlelike micropores between the graphitic fibrils. The average diameters of these pores depend
on the heat treatment of the fiber, but lie typically in the range 6 to 20 Å.
Although these micropores may account for up to 30% of the total porosity,
they are too small to be detected by nitrogen-adsorption measurements.

B. Polyacrylonitrile-Based Fibers

Polyacrylonitrile-based fibers generally show somewhat lower surface
areas than those derived from a-cellulose or rayon, typical values being in
the range 0.1 to 0.4 m^2/g for Modmor fibers and 0.5 to 1.5 m^2/g for the
Thornel series. This difference is consistent with the general morphology
of the two types of fiber, those prepared from PAN having a uniform circular
cross section and very smooth surfaces, whereas those from cellulosic precursors show irregular cross sections and rough fluted surfaces (see Section III.B).

A distinction should be made between external and internal surface
areas, the latter being quite large because of the presence of internal pores
and voids that lie along the axis of the fibers but do not extend to the surface.
Thus D.J. Johnson and Tyson [6] interpret the results of low-angle X-ray
diffraction studies on PAN-based fibers in terms of stacks of short graphitic
crystallites separated by sharp-edged voids, the mean width of which is less
than 10 Å. Although the calculated surface area of powdered fiber samples
is on the order of 190 m^2/g, much of this area is not accessible to gases
adsorbed on whole fibers.

A detailed study of the effects of heat treatment on the surface areas of
Modmor fibers has recently been made in the authors' laboratory [39]. A
sample of chopped Modmor II fibers (surface treated) was sealed into a
sensitive volumetric adsorption apparatus and evacuated to 2 x 10^{-7} torr
at room temperature. The surface area, subsequently measured by krypton
adsorption at -195°C, was found to be 0.24 m^2/g. The fiber sample was
then evacuated at a series of increasing temperatures 100°C apart, the
krypton surface area being measured after each evacuation. After each
increment in the evacuation temperature additional gas was desorbed from

the fibers. The composition of this desorbed gas was found by in situ mass-spectrometry analysis to be a complex mixture of the oxides of carbon and nitrogen, water, methane, hydrogen, and C_3 and C_4 unsaturated hydrocarbons. The concentrations of desorbed hydrocarbons decreased with increasing temperature, and above $500^{\circ}C$ only carbon monoxide, carbon dioxide, and hydrogen were detected. With increasing degassing temperature the surface area of the Modmor II fibers increased continuously, as shown by curve A in Fig. 25, the value after a $1000^{\circ}C$ bakeout corresponding to a fourfold

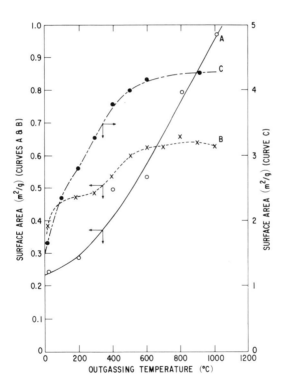

FIG. 25. Surface areas of Modmor fibers as a function of outgassing temperature [39]. Curve A ($\circ$), Modmor II, treated; curve B (x), Modmor I, untreated; curve C ($\bullet$), Modmor I, after reflux with 70% HNO_3 for 72 h.

increase over that of the original fibers at room temperature. Separate measurements with fibers that had been extracted with xylene for 48 h showed the same variation in surface area with degassing temperature, indicating that the effect was not due to carbonization of an extractable surface additive introduced by the manufacturer.

Corresponding data for a Modmor I untreated fiber sample are shown by curve B of Fig. 25. In this case the increase in surface area was much less than it was in Modmor II, the area reaching a constant value of about 0.62 m^2/g at outgassing temperatures of $600^{\circ}C$ and above. It seems likely that the difference in the behavior of the two types of fiber is related to their structure, the more highly graphitized Modmor I having already been processed to temperatures in excess of $2500^{\circ}C$ during manufacture. In the case of Modmor II it is likely that heat treatment results in desorption of residual precursor decomposition products from pores already present in the structure.

The effect of an initial oxidative treatment on the subsequent development of surface area in Modmor I is shown in curve C of Fig. 25. After refluxing with strong nitric acid for 72 h the room-temperature surface area of the fibers was increased by a factor of 4, and subsequent heat treatment resulted in a much larger increase in surface area than was observed with the untreated fiber. As discussed in Section V, oxidation with nitric acid appears to increase the surface area by micro-etch-pit formation. Other treatments, such as oxidation by gaseous oxygen or carbon dioxide, would also be expected to result in substantial increases in surface area and porosity, as with crystalline graphite [59]. The effects of these surface-area changes on the mechanical properties of the fibers are discussed in Section V.

V. SURFACE TREATMENT OF CARBON FIBERS

In the early stages of the development of high-strength carbon-fiber composite materials a severe limitation was found to be the weak bonding between fiber and matrix and the resulting low interlaminar shear strength of the composite. In this respect carbon fibers were distinctly inferior to other fibers used in structural materials, the shear strengths of graphite-fiber composites being on the order of 4000 psi, compared with about 15,000 psi for boron and S-glass filament composites. In general, as the degree of graphitization and modulus of carbon fibers were increased, the composite shear strength progressively decreased, as shown in Fig. 26. This unfortunate situation has to a large extent been remedied by the development of special fiber-surface treatments that improve the bond between the reinforcing fiber and the matrix material. Although several theories have been proposed, the effects on composite properties produced by fiber-surface treatments are not well understood.

Although vacuum impregnation of carbon fibers with resin results in some improvement in fiber-matrix bonding, as indicated by composite mechanical properties [60], most coatings are applied to surface-treated fibers to improve their handling and protect the surface from abrasion. The most common surface-treatment methods used in the past are listed in Table 5.

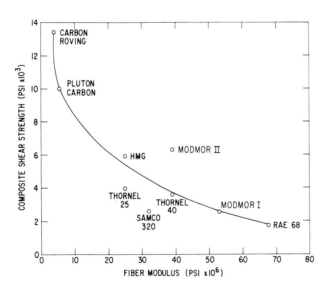

FIG. 26. Shear strengths of carbon fiber-epoxy composites as a function of fiber modulus. After Goan and Prosen [57].

TABLE 5

Surface Treatment of Carbon Fibers

Oxidative etching

Wet methods	Dry methods
Reagents:	Reagents: Vacuum desorption
Nitric acid	Air
Permanganate-sulfuric acid	Oxygen
Chromic acid and dichromates	Ozone
Hypochlorous acid	Catalytic oxidation
Sodium hypochlorite	

Coatings

Polymers

Pyrolytic graphite

Whiskerizing (silicon carbide)

Inorganic coatings (e.g., silica, metals)

A. Oxidative Etching--Wet Methods

1. Nitric Acid Treatment

a. Rayon-Based Fibers. The most commonly used and the most consistently beneficial surface treatment has been a wet oxidation method using
nitric acid as the oxidant. A detailed discussion of nitric acid surface treatments has been given recently by Herrick and co-workers [41, 42]. They
used a graphitic yarn (WYB 85-1/2) and reported an increase in surface area
from 1 to 136 m^2/gm and a 12% weight loss after refluxing for 24 h in 60%
nitric acid. This treatment also resulted in increased carboxyl-group concentrations as estimated from calcium acetate neutralization values (i.e.,
the liberated acetic acid was titrated with dilute potassium hydroxide, as
described in Section VI). More dilute acid was ineffective in increasing the
surface functionality, as shown in Fig. 27. The changes in surface functionality and surface area of WYB 85-1/2 graphitic yarn and Thornel 25 fiber
after various acid treatments are summarized in Table 6.

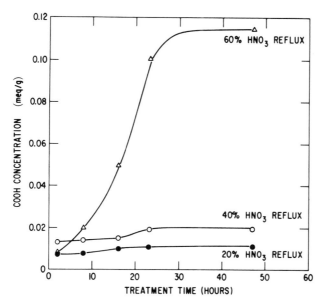

FIG. 27. Development of surface functionality in graphitic yarn (WYB 85-1/2) during HNO_3 treatment. After Herrick et al. [42].

TABLE 6

Changes in the Surface Properties of Carbon Fibers
after Nitric Acid Treatment[a]

Fiber	Treatment	Functionality (meq/g)	Surface area (m^2/g)
WYB 85-1/2	None	--	2
	60% HNO_3 + 16-h water wash	0.25	338
	60% HNO_3 + 3 water boils	0.02	32
	70% HNO_3 + 16 water boils	0.10	126
Thornel 25	None	0.002	1
	60% HNO_3 + 3 water boils	0.005	6
	70% HNO_3 + 3 water boils	0.006	12

[a]Data from Ref. [42].

An attempt to distinguish between different types of functional groups introduced by the acid treatment was also made. Table 7 shows the neutralization values of Thornel 25 after treatment with 70% nitric acid, as determined by using three bases. As discussed in Section VI, calcium acetate gives a general test for carboxyl groups, sodium bicarbonate reacts with all acid groups except phenolic hydroxyl, and sodium hydroxide measures the total acid-group concentration. These results apparently indicate that a large concentration of phenolic hydroxyl groups are present on the fiber surface.

TABLE 7

Surface Groups on Thornel 25 after Nitric Acid Treatment[a]

Base	Groups measured	Surface activity (meq/g)
Calcium acetate	Carboxyl	0.04
Sodium bicarbonate	Carboxyl, lactone, other acid groups	0.32
Sodium hydroxide	All acid groups, phenolic hydroxyl	4.44

[a]Data from Ref. [42].

Herrick's results [42] on the effects of nitric acid treatment of the fibers on the mechanical properties of carbon fiber-epoxy composites are summarized in Table 8. In these tests the acid treatments were conducted at boiling temperature ($\sim 120^\circ$C) for 24 h. The data indicate that, although the nitric acid treatment resulted in a marked increase in shear strength, the tensile strengths were often reduced. This weakening of the fibers was especially noticeable after treatment with 70% acid.

Although oxidation of graphite fibers with nitric acid increases both the surface area and chemical functionality, Herrick [42] showed that the chemical activity of the fibers could be eliminated by heating in a hydrogen furnace at 500°C without effecting the surface area. Composites made from fibers treated in this way exhibited shear strengths near the values obtained by using

TABLE 8

Effect of Nitric Acid Treatment on the Mechanical Properties
of Carbon Fiber-Epoxy Composites[a]

Fiber	Treatment	Shear strength			Longitudinal tensile strength (psi)
		Flex.	Core	Torsion	
WYB 85-1/2	None	3300	2400	3900	50,000
	60% HNO_3	>6200	3700	7300	46,000
	70% HNO_3	3400	3900	--	28,000
Thornel 25	None	4000	2500	3800	80,000
	60% HNO_3	5000	2500	--	95,000
	70% HNO_3	>6000	3500	6000	50,000

[a]From Ref. [42].

untreated fibers. As hydrogen reduction under these conditions is known to decompose surface carboxyl groups, Herrick concluded that surface functionality is more important than surface area in determining composite shear strength.

This may not be a general conclusion. More recent work by Scola and Brooks [58] on rayon-based Hitco fibers is in direct conflict with Herrick's results. As shown in Table 9, hydrogen treatment of the oxidized fibers caused a further increase in surface area and no decrease in shear strength.

Scola and Brooks [58] point out that as the surface area is increased, the carboxyl-group concentration per unit area may only increase slightly or even decrease, so that it becomes very difficult to assess the relative importance of surface area and surface acidity.

This point is illustrated further by the data in Table 10, in which the adsorption properties of Hitco graphitic yarn after successive oxidation treatments with nitric acid are correlated with changes in the mechanical properties of the resulting fiber-epoxy composite. Although increasing oxidation is accompanied by a steady increase in fiber-surface area and adsorption of bases, the concentration of acid sites per unit area decreases

TABLE 9

Effect of Surface Treatment on Graphite Composites
Derived from Hitco HMG-50[a]

Treatment	Fiber surface area (m^2/g)	Composite shear strength (psi)
None	0.87	4400
Oxidation	5.34	5930
Oxidation, then H_2 at $600^{\circ}C$ and atmospheric pressure for 3 h	8.75	5870

[a]Data from Ref. [58].

TABLE 10

Surface Properties and Composite Shear Strengths of Hitco HMG-50
Graphitic Yarn and HMG-50-Epoxy Resin Composites[a]

Treatment	Specific surface area (m^2/g)	NaOH adsorbed $(\mu mole/g)$	NaOH adsorbed $(\mu mole/m^2)$	Short-beam shear strength (psi)	Transverse strength (psi)
None	0.87	0.066	0.074	4490	--
Oxidation	3.4	2.04	0.6	5600	1800
Oxidation	7.3	3.0	0.41	6090	2790
Oxidation	10.5	3.60	0.34	6500	3080
Oxidation	24.0	5.9	0.25	7000	--

[a]From Ref. [54].

gradually as the surface area rises from 3.4 to 24 m^2/g. Both the shear strength and transverse strength of the composite increase with increasing oxidation of the fibers.

The changes in surface morphology that accompany the treatment of the carbon fibers with nitric acid have not been studied in detail, although Herrick [42] reports that gross surface roughness is reduced in rayon-based fibers after reflux with nitric acid. Increase in surface area probably results from the formation of micro etch pits on the surfaces of the fibers, analogous to those observed on the basal plane surfaces of crystalline graphite after oxidation [61]. Refluxing natural graphite crystals with 70% nitric acid for 72 h resulted in considerable swelling and exfoliation of the layers, and, in addition, the basal plane surfaces exhibited the mottled appearance shown in Fig. 28 (right). Such micropitting, which may be of molecular dimensions, may well be responsible for the surface-area increase found with carbon fibers. Also, by exposing additional prismatic surfaces, formation of etch pits can result in increased concentrations of polar groups at reactive edge-carbon sites. It is quite likely, however, that the pores introduced by nitric acid treatment are small enough to exclude resin molecules while admitting the nitrogen or krypton used in surface-area determinations. In this case the composite shear strength should be independent of the surface area, a result that has been observed in the case of PAN-based fibers as discussed below. In the general case both surface area and surface functionality can be expected to influence composite shear properties, the relative importance of each factor depending on the detailed structure of the fiber surface.

b. **Polyacrylonitrile-Based Fibers.** Fibers derived from polyacrylonitrile appear to be more readily oxidized by nitric acid treatment than are fibers based on rayon [38]. A detailed study of the effect of nitric acid treatment on the properties of Modmor fibers has recently been made by the authors. Modmor I (high modulus, low strength) and Modmor II (low modulus, high strength) were refluxed in 60% nitric acid for varying periods. After reaction the fibers were removed from the acid, rinsed in distilled water, rinsed four times in boiled distilled water, and then flushed with distilled water. The fibers were air-dried at $110^{\circ}C$ and then vacuum-dried at $50^{\circ}C$ for several hours. The fibers were weighed before and after the acid treatment. Small, uniaxial, fiber-reinforced composite bars (1.5 x 0.08 x 0.05 in.) were then made, using the oxidized fibers as the reinforcement and

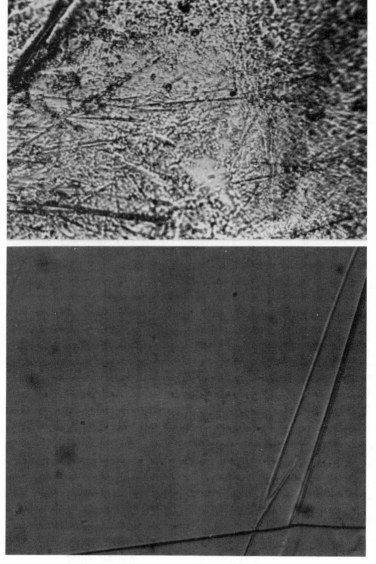

FIG. 28. Changes in the topography of basal plane surface of natural graphite crystals during nitric acid treatment. Left: typical cleaved Ticonderoga graphite basal plane with twin bands (x384); right: appearance of basal plane surface after reflux with 70% HNO_3 for 72 h (x384).

epoxy as the matrix. The flexural strength and interlaminar shear strength
of the composite bars were measured in a three-point flexure test, the
span-to-depth ratio being 24 for flexural strength and 6 for interlaminar
shear strength measurements.

It was found that PAN-based fibers vary widely in their susceptibility
to surface activation by nitric acid depending on the temperature at which the
fibers have been carbonized or graphitized. Modmor I fibers have been
graphitized at ~2800°C and are characterized by graphitic crystallites, about
200 Å in length, which are highly oriented with respect to the fiber axis.
This fiber lost 4% of its weight in boiling 60% nitric acid during the first 3 h,
but was unaffected by further treatment, as shown in Fig. 29. On the other
hand, Modmor II fibers have been heat-treated to about 1200°C and are
characterized by very small graphitic "nuclei" with a higher degree of misorientation of the "nuclei" with respect to the fiber axis [21]. This fiber
lost 50% of its weight after 72 h in boiling 60% nitric acid, with smaller
weight losses at intermediate treatment times (Fig. 29).

No microscopically observable differences were found between the untreated and acid-treated Modmor I fibers at x5000 magnification in the scanning electron microscope, whereas several significant differences were
observed between the untreated and acid-treated Modmor II fibers. The

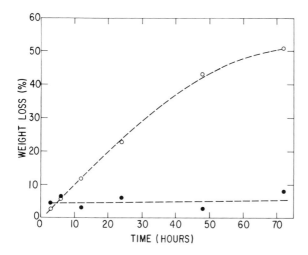

FIG. 29. Weight loss of Modmor I fibers (●) and Modmor II fibers (o)
as a function of time in 60% HNO₃ at 118°C.

diameter of the fiber was reduced from about 8 μm to about 5.5 μm, and the microscopic roughness (i.e., striations) were removed (Fig. 30). In addition, the acid solution after treatment of type II fibers contained small black particles in suspension which eventually settled to the bottom of the reaction vessel, whereas the solution from the type I fibers remained clear.

The effect of nitric acid treatment of Modmor type I and II fibers on composite mechanical properties is shown in Figs. 31 through 33. The results show a 30% increase in flexural strength for acid-treated type I fibers in an epoxy matrix but a 43% decrease for acid-treated type II fibers. The interlaminar shear strength increased by a factor of 3 to 10,000 psi for acid-treated type I fibers, whereas only a slight improvement (~25%) was obtained in the shear strength of composites reinforced with acid-treated type II fibers, with the best results obtained after short treatment times.

Although it is not obvious from these results whether the improved shear properties are due to increased surface area or increased surface functionality, further experiments have shown that outgassing these oxidized fibers in vacuum at temperatures of up to 900°C causes substantial increases

FIG. 30. Scanning electron micrographs of Modmor II fibers before (left) and after (right) reflux with 60% HNO_3 for 72 h. x 5000.

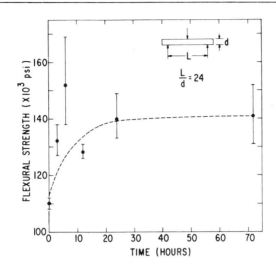

FIG. 31.   Effect of nitric acid oxidation time on the flexural strength of composites reinforced with treated Modmor I fibers.

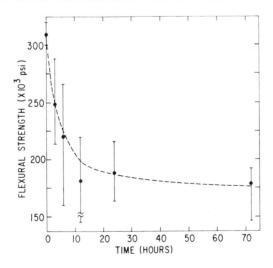

FIG. 32.   Effect of nitric acid oxidation time on the flexural strength of composites reinforced with treated Modmor II fibers.

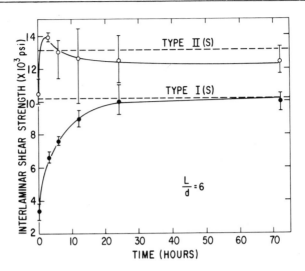

FIG. 33. Effect of nitric acid oxidation time on the interlaminar strength of composites reinforced with treated Modmor fibers.

in surface area (see Section IV) but very little change in composite shear strength. As it is highly likely that acid groups like carboxyl would be decomposed completely under these conditions, it is possible that the loss of surface functionality is compensated by the increase in surface area. Scola and Brooks [58] found a correlation between the specific surface area of cellulose-based fibers and composite short-beam shear strength but observed no such correlation for acrylic-based fibers. Data for the changes in surface properties and composite shear strengths of oxidized Modmor fibers on outgassing are summarized in Table 11.

The lack of correlation between composite shear strength and surface area and/or surface functionality due to oxidative treatments still leaves unexplained the nature of the improvement. It is possible that the improved shear properties result not so much from increasing the interaction between matrix and reinforcement but from removing from the surface of the reinforcement "defects" that act as stress risers, causing failure to occur at less than optimum values.

TABLE 11

Effect of Fiber-Surface Area on Composite Shear Properties

Fiber[a]	Treatment	Surface area (m^2/gm)	Interlaminar shear strength[b, c] $(10^3$ psi$)$
Type I	None	0.38	3.3
	Outgassed at $900^{\circ}C$	0.62	4.1
	60% HNO_3 72 h at $118^{\circ}C$	1.64	10.1
	60% HNO_3 72 h at $118^{\circ}C$, outgassed at $900^{\circ}C$	4.26	10.0
Type II	Supplier's treatment	0.24	13.1
	Supplier's treatment, outgassed at $900^{\circ}C$	0.92	12.2

[a]Fibers were obtained from Morganite Modmor Ltd. Type I is high-modulus, low-strength; type II is low-modulus, high-strength.

[b]Short-beam shear, $L/d = 6$; epoxy matrix.

[c]All values are averages of at least three samples.

Stereoscan micrographs of fracture surfaces of composites reinforced with untreated and treated Modmor fibers (Fig. 34) show the poor interfacial bonding of untreated Modmor I - epoxy composite after short-beam shear testing (ILS = 3500 psi). Fiber pullout is obvious as well as actual fiber-from-resin separation. The treated type I fibers, shown in micrographs B in Figs. 34a and 34b, show less fiber pullout and good wetting of fiber by the resin. The difference in the appearance of the fracture surfaces in micrographs B and C is apparently due to the mode of failure, that is, primarily shear failure in fibers treated by supplier (micrographs B) and tensile failure in acid-treated type I fibers (micrographs C). Similar results were obtained with Modmor type II fibers.

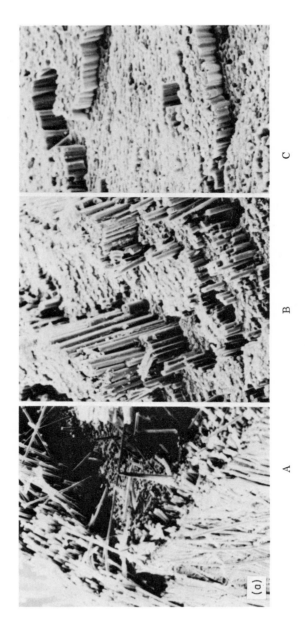

FIG. 34a. Scanning electron micrographs of fracture surfaces of epoxy composites reinforced with Modmor I fibers: A — untreated fibers (x100); B — fibers treated by supplier (x200); C — fibers oxidized in 60% HNO_3 for 72 h at 118°C (x200).

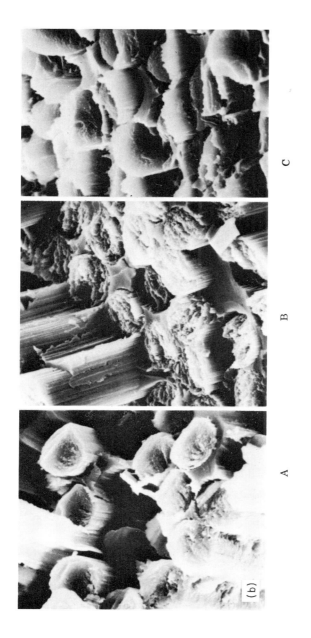

FIG. 34b. Scanning electron micrographs of the fracture surfaces shown in Fig. 34a at increased magnification (×2000): A — untreated fibers; B — fibers treated by supplier; C — fibers oxidized in 60% HNO_3 for 72 h at 118°C.

2. Other Wet Oxidation Treatments

Various other aqueous oxidizing solutions have been used to treat carbon-fiber surfaces. Bobka and Lowell [34] used a $KMnO_4$-H_2SO_4 solution to etch graphitic yarn, the extent of etching being controlled by varying both solution temperature and immersion time. The results of these experiments are shown in Table 12.

TABLE 12

Properties of Etched Graphitic Yarn-Resin Composites[a]

Etching conditions		Surface area of yarn (m^2/g)	Flexural strength of composite $(10^3$ psi)	Compressive strength of composite $(10^3$ psi)
Solution temperature ($^{\circ}$C)	Immersion time (min)			
Unetched yarn	--	4	71	63
50-55	5	4	57	124
	10	5	47	134
95-105	5	8	35	105
	10	20	37	136
	20	12	28	--
	24	23	28	--

[a]From Ref. [34].

Increased etching was accompanied by increasing the surface area and compressive strength of the yarn, although the composite flexural strength decreased continuously. By optimizing the etching conditions it was possible to double the composite shear and compressive strengths with only small decreases in tensile and flexural strengths. According to Herrick et al. [42], for treatments of short duration, permanganate solutions are more effective then nitric acid in introducing carboxyl groups into rayon-based carbon yarn.

Treatment of Thornel 25 fiber with sodium hypochlorite solution has been shown by Scola and Brooks [58] to result in modest increases (about 30%) in composite shear strength, and hypochlorites have also been claimed to inhibit excessive etch-pit formation during air oxidation of fibers [62].

That not all strong oxidizing agents are effective in enhancing the properties of carbon fibers is shown by the work of Didchenko [37], who treated Thornel fibers with chromic acid cleaning solution. As shown in Table 13, treatment for 5 min at $50^{o}C$ with this reagent resulted in a loss of surface area and a sharp increase in pores of large diameter. These changes were accompanied by a marked decrease in tensile strength as a result of excessive degradation of the fibers.

TABLE 13

Change of Porosity and Surface Area of Thornel 40 after
Treatment with Chromic Acid Cleaning Solution[a]

Sample	Surface area (m^2/g)	Pore volume (ml/g)	Median pore diameter $(\overset{o}{A})$
Untreated Thornel 40	1.4	0.00306	144
After treatment with cleaning solution at $50^{o}C$ for 5 min	0.3	0.00251	475

[a]From Ref. [37].

B. Oxidative Etching--Dry Methods

1. Vacuum Desorption

Some improvement in fiber properties is obtained by evacuating carbon fibers at elevated temperatures before impregnation with the resin. Thus Herrick, Gruber, and Mansur [42] increased the composite shear strength of graphitic yarn by about 20% by degassing the yarn at $300^{o}C$ and 1 torr for several minutes before vacuum impregnation with the epoxy resin. This treatment probably desorbs impurities from the fiber surface and thereby promotes a strong bond between fiber and resin. The effect is, however, minor compared with those produced by chemical etching. Outgassing of acrylic-based carbon fibers in vacuum ($<10^{-5}$ torr) also has a minor effect on the composite shear properties, increasing the ILS from about 3500 psi to 4100 psi [63].

2. Air Oxidation

Many attempts have been made to develop an air-oxidation method for carbon-fiber surface treatment. In spite of some success, the process is difficult to control, and results are often unreproducible. Herrick, Gruber, and Mansur [42] report little change in the mechanical properties of rayon-based graphitic yarn after heating in air at 500°C for 16 h, although treatment for the same period at 600°C gave a 45% increase in shear strength. Thornel fibers, after air oxidation at 400°C for 16 to 30 h, showed an increase in composite shear strength from 4100 to 5900 psi for Thornel 25 and to 5600 psi for Thornel 40 [57]. The increase in surface area and surface functional groups as a result of air oxidation of Thornel 40 is shown in Table 14. An

TABLE 14

Effect of Air Oxidation on Thornel 40 Surface Properties[a]

		Surface groups (%)		
Sample	Surface area (m^2/g)	H	COOH	Phenolic OH
Untreated Thornel 40	0.84	0.01	·0.11	0.13
Air oxidized at 400°C	2.0	0.03	--	0.4

[a]From Ref. [57].

increase in surface carboxyl-group concentration on graphitic yarn after air oxidation at 550°C has also been reported by Herrick et al. [42].

In a recent study of factors influencing the mechanical strength of PAN-based fibers, J.W. Johnson [27] concludes that an important effect of mild air oxidation is to remove structural flaws and cavities that are responsible for tensile failure in untreated fibers. A 10-min treatment in air at 450°C was found to be most effective and gave the statistical distribution of breaking stress in the untreated and oxidized fibers shown in Fig. 35.

In spite of these reported improvements, the conditions used during air oxidation are critical in determining the ultimate composite shear properties, excess oxidation producing marked pitting of the fiber surface and reduced tensile strength [64]. This result is to be expected, as the oxidation of crystalline graphite is known to result in etch-pit formation on the basal

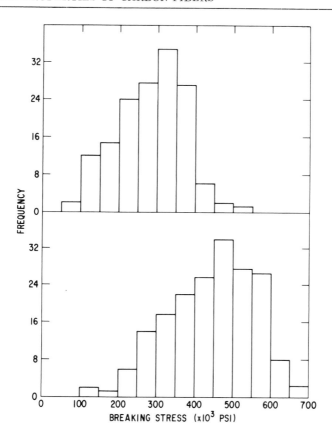

FIG. 35. Strengths of PAN-based fibers (Courtelle): Top: untreated; bottom: after oxidation in air at $450^{\circ}C$ for 10 min. After J.W. Johnson [27].

plane surfaces [61]. Such excessive pitting and loss of mechanical strength has been reported for Modmor I fibers after air oxidation at $550^{\circ}C$ for 16 h by Harris, Beaumont, and Rosen [26]. The pitted surface of these fibers is shown in Fig. 36 for 5- and 16-h exposure, respectively.

Further details of the oxidative erosion of Modmor I and II fibers have recently been observed by us. We used a thermogravimetric balance to follow the weight changes of the fiber sample during oxidation in a stream of air at $700^{\circ}C$. The oxidation was terminated at various percent burnoff values, and the surfaces of the residual fibers were examined by means of

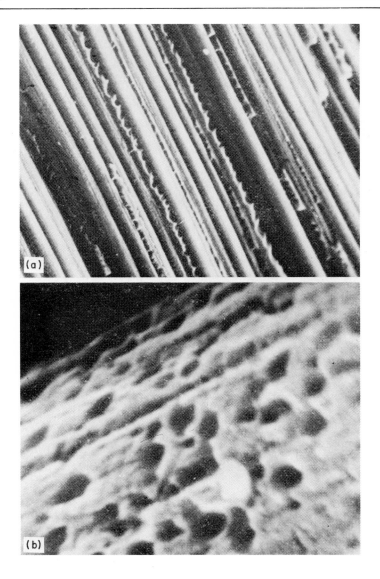

FIG. 36. Modmor I fiber surface after oxidation in air. Top: after heating for 5 h at 550°C (x280); bottom: after heating for 16 h at 550°C (x1500). Reprinted by permission from Ref. [26].

a scanning electron microscope. Figure 37a shows that after 8% weight loss, the surface of the fibers is quite uniformly pitted by arrays of elliptical etch pits. A marked amount of powdered material is also in evidence. Figure 37b shows the appearance of the fibers after oxidation in air to 30% weight loss.

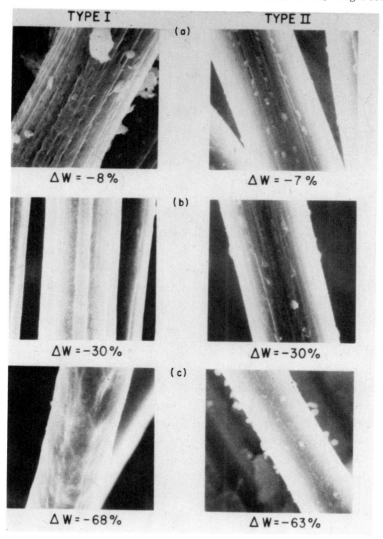

FIG. 37. Surface topography of Modmor fibers after air oxidation: (a) after ~8% burnoff at $700^{\circ}C$; (b) after 30% burnoff; (c) after ~65% burnoff. x 3500

Pitting is not now observable, but the macroscopic striations parallel to the fiber axis that were initially present on the fiber surface are still evident. Figure 37c shows the appearance of the fibers after oxidation in air to 68% weight loss. Although the mean diameter of the fibers has been reduced from 8 to 6 μm, the surface topography is now more uniform, all trace of etch pits having disappeared. In addition all powdered material has been removed. It seems likely that etch pits are only produced in the early stages of oxidation, continued burnoff resulting in the removal of concentric layers of carbon. The fibers in Fig. 37c do, however, show a somewhat variable diameter, indicating that the oxidation rate may not be quite uniform perpendicular to the fiber axis. To minimize the effects of excessive burnoff, Wadsworth and Watt [65] suggest that the weight loss of the fibers during oxidative surface treatment be kept below 6 wt %.

It has recently been claimed [66] that sulfur dioxide, halogens, and halogenated hydrocarbons are effective in inhibiting excessive pitting of fibers during oxidative surface treatment by air or oxygen, although we have failed to observe any inhibitory effect produced by chloroform or carbon tetrachloride on the oxidation behavior of Modmor fibers or crystalline graphite [63]. A similar effect has also been reported for hypochlorite solutions [62] produced electrochemically by using the carbon fibers as an anode in an aqueous NaCl-NaOH solution. Phosphorus compounds are well known as inhibitors for graphite oxidation [67], but apparently have not yet been used in carbon-fiber treatment.

3. Oxidation with Gaseous Oxygen and Ozone

For obvious reasons surface oxidation of carbon fibers by means of gaseous oxygen or ozone is even more difficult to control than the air process. Temperatures in excess of 400°C are required to bring about the reaction in the case of oxygen, whereas above 600°C spontaneous ignition of the fibers often occurs. The oxidation characteristics of various types of carbon fiber are discussed in Section VII.

Early work by Bobka and Lowell [34] on the oxidation of graphitic yarn with flowing oxygen at 400 to 600°C showed that etching of the fibers was very nonuniform, about 5% of the fibers being greatly degraded and the rest remaining unoxidized.

Goan and Prosen [57] have reported the results of various oxidizing treatments on Thornel 25 including exposure of the fibers to reduced

pressures of oxygen in the presence of an rf discharge and heating in the
presence of an oxygen–ozone mixture. The results are summarized in
Table 15. The effect of ozone was less dramatic than expected in view of
the reported vigorous attack of ozone on crystalline graphite [68].

TABLE 15

Effects of Dry Oxidation on Composite Shear
Strength of Thornel 25[a]

Fiber treatment	Average shear strength (psi)
Untreated	4100
Oxidized in rf discharge	5800
Oxidized in O_2-O_3 mixture	5000

[a]Data from Ref. [57].

4. Catalytic Oxidation

We have recently developed a surface-treatment technique for carbon
fibers that utilizes the catalytic properties of certain metal oxides. The
oxidation of graphite is greatly accelerated by small amounts of metallic
impurities, especially the oxides of copper, lead, vanadium, and the transition metals [69]. Particles of these catalysts often show curious local effects when placed on the basal-plane surface of a graphite crystal heated in
oxygen to temperatures of 300 to 700°C. Thus oxidation in the presence of
copper oxide is accompanied by rapid movement of the catalyst particles
on the graphite surface and the formation of a network of shallow irregular
channels [70]. Attack in the c-axis direction appears to be slow in this
case. Other catalysts, such as lead, manganese, and molybdenum oxides,
promote reaction in the c-axis direction and lead to rapid pitting of the
entire graphite surface at temperatures considerably below those required
for uncatalyzed oxidation. Typical catalytic channeling and pitting on natural
graphite crystals are illustrated in Figs. 38 and 39. It seems possible that
controlled catalytic oxidation of this type can be used to roughen the surfaces
of carbon fibers without weakening the fibers by extensive etch-pit formation.

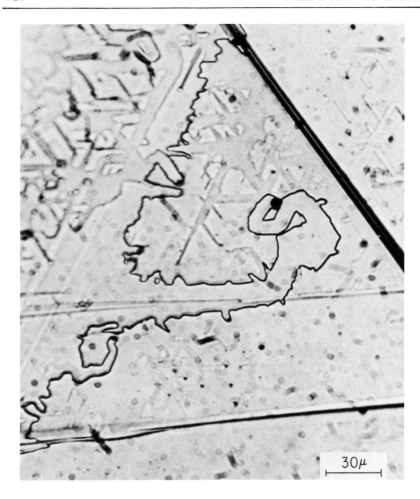

FIG. 38. Catalytic channeling by vanadium oxide particles on natural-graphite basal plane during oxidation at $600^{\circ}C$ in oxygen. Reprinted by permission from Ref. [69].

Four tows of Modmor I fibers were catalytically treated as follows:

Tow A. The fiber tow was immersed for 15 min in hot 0.1% cupric acetate solution. After drying in air, the sample was heated in air to $500^{\circ}C$ for 30 min. After cooling, the tow was washed first with dilute acetic acid and then with distilled water, and dried in air.

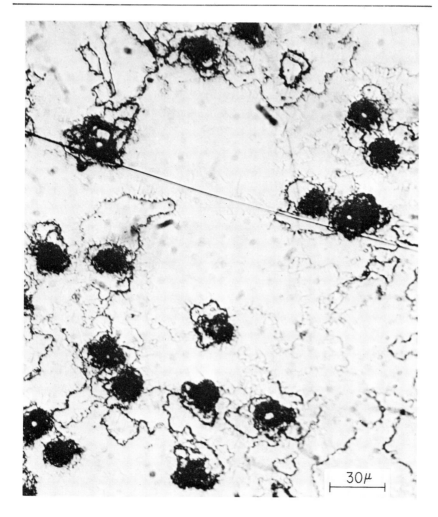

FIG. 39. Catalytic pitting by molybdenum oxide particles on natural-graphite basal plane during oxidation at 700°C in oxygen.

Tow B. The tow was treated with copper acetate solution as before, heated in flowing oxygen in a tube furnace to 600°C for 10 min, and then washed and dried as before.

Tow C. The tow was immersed for 15 min in hot 0.1% lead acetate solution. After drying in air, the sample was heated in air to 500°C. After

5 min the tow became red hot, and extensive degradation occurred. The temperature used in this case was apparently excessive, as lead is a more active catalyst for graphite oxidation than copper [69].

Tow D. The tow was treated with lead acetate solution as before, then heated in air to 400°C for 30 min, and finally washed and dried as before.

The mechanical properties of composites prepared from fiber samples A, B, and D and the values for untreated Modmor I are summarized in Table 16.

TABLE 16

Catalytic Oxidation of Modmor I Fibers

Sample	Flexural strength (10^3 psi)	Flexural modulus (10^6 psi)	Interlaminar shear (10^3 psi)
Untreated	110	28	3-4
A (Cu treated)	84	29.9	6.81
	80	25.7	6.20
	89	28.0	6.46
B (Cu treated)	118	30.0	4.38
	114	30.0	4.94
	104	27.1	5.00
D (Pb treated)	118	28.7	4.27
	121	30.4	4.75
	116	28.0	4.09

It appears from these results that mild catalytic oxidation of carbon fibers can result in substantial improvements in composite shear properties. The process will, however, need to be carefully optimized to avoid excessive degradation.

C. Surface Coatings

1. Polymers

Much attention has been paid to the possibility of increasing the fiber-resin interfacial bond by the application of polymers or reactive monomers to the fiber surface, which is usually given a preliminary oxidation treatment by means of nitric acid. Polymeric materials, such as polyvinyl alcohol, are also often added as "sizing" to improve the handling character-istics of the fibers, and many commercial fibers having a coating of this sort.

An extensive study of the effects of polymer coatings on the mechanical properties of epoxy-graphitic yarn composites was carried out by Herrick [71]. Before application of the polymer the fibers were oxidized with 60% nitric acid for 24 h. The effect of various coatings on the composite shear strength is summarized in Table 17.

TABLE 17

Effect of Polymer Coatings on Composite Shear Strength[a]

Polymer coating	Percent polymer	Density (g/ml)	Core shear strength (psi)
None[b]	--	1.28	2350
None	--	1.29	3520
Polyvinyl alcohol	7	1.31	6200
Polyvinyl chloride	7	1.31	6100
Rigid polyurethane	3	1.27	5900
Polyacrylonitrile	7	1.27	2400

[a]Data from Ref. [71]; WYB 85-1/2 fiber. Unless otherwise specified, fibers were given a preliminary 24-h oxidation treatment in 60% HNO_3.
[b]Untreated sample.

Although there is marked improvement in shear strength in many cases, flexure tests on these samples generally led to tensile failure, which became more frequent when 70% nitric acid was used in the preliminary oxidation. Similar improvements in composite shear properties were obtained by coat-ing oxidized Thornel 25 fibers with polyvinyl chloride and urethane, although in this case 70% nitric acid was found to be the most effective oxidant. Both Herrick [71] and Goan and Prosen [57] have used polyamide as a coating ma-terial for carbon fibers, with little effect on the shear strengths of composites.

Although organosilanes have shown considerable success as coating agents for glass fibers, their application to carbon fibers has been rather disappointing. Coating of unoxidized rayon-based carbon fibers with chloro and alkoxy silanes had a somewhat adverse effect on composite shear properties [60], although with fibers preoxidized with nitric acid some slight improvement in shear strength was observed by Goan and Prosen [57] when γ-aminopropyl triethoxysilane was used as the coating agent. It seems possible that the oxidation treatment produces surface polar groups that interact with the silanol groups of the coating.

A detailed study of silane coatings on Modmor I carbon fibers has been made recently by Harris, Beaumont, and Rosen [26]. The fibers were first boiled with 70% nitric acid for 2 h and then treated with 30 vol % solutions of trialkoxy silanes in methyl ethyl ketone solvent. After immersion for 1 h the weight of the dried fibers had increased by about 15%, corresponding to a mean coating thickness of 0.5 μm. The effects of the coatings on the composite mechanical properties are summarized in Table 18.

In the most favorable case (silane A186) a total improvement of about 35% in shear strength was obtained over that of the untreated fiber. However, most of this increase in shear strength could be attributed to the effects of the nitric acid oxidation, and the other silanes appeared to give very marginal improvement in composite properties. On the whole there is little evidence that dramatic changes in fiber-surface properties can be induced by coatings of this type.

2. Vapor-Phase-Deposited Coatings

Many materials have been deposited on carbon fibers from the vapor phase, the intention being either to increase the shear properties of the fibers or to improve their oxidation resistance. Schmidt and Hawkins [46] succeeded in depositing pyrolytic graphite on graphite yarn by passing methane through fiber layers at 1450 to 1500°C. Hydrogen added as a diluent to the methane stream served to control the rate of deposition of the graphite coating. In addition it was found that pyrolytic graphite doped with various metals (e.g., B, Ti, Hf, and Zr) could be deposited by premixing the methane with a volatile halide of the metal. These coatings, together with pyrolytic nitrides and carbides, were found to increase the oxidation resistance of the yarn.

TABLE 18

Effect of HNO_3 Treatment on the Mechanical
Properties of Modmor[a] I

Fiber treatment ($V_f = 0.40$)	Tensile failure strength (kg/mm^2)	Flexural strength (kg/mm^2)	Interlaminar shear strength (kg/mm^2)	Ratio of shear strength to tensile strength
As-received	52.8	38.4	2.06	0.039
Boiled 2 h in conc. HNO_3	56.6	47.5	2.52	0.045
Boiled in HNO_3 + A186[b]	63.7	53.0	2.77	0.044
Boiled in HNO_3 + A187[c]	62.0	56.8	2.66	0.043
Boiled in HNO_3 + A189[d] (1)	39.5	24.8	1.82	0.046
Boiled in HNO_3 + A189[d] (2)	23.5	18.3	2.58	0.11
Heated in air at 550°C for 5 h	26.6	49.5	Tensile failure at 28.8	--
Heated in air at 550°C for 16 h	20.7	39.0	Tensile failure at 26.1	--

[a]Data from Ref. [26].
[b]Union Carbide Corp. code for β-(3,4--epoxycyclohexyl-)ethyl-Si-$(OCH_3)_3$.
[c]Union Carbide Corp. code for γ-glycidoxypropyl-Si-$(OCH_3)_3$.
[d]Union Carbide Corp. code for mercaptopropyl-Si-$(OCH_3)_3$.

Herrick et al. [42] and Goan and Prosen [57] have also succeeded in depositing a variety of coatings on Thornel carbon fibers. Among the most interesting were silica (from the pyrolytic decomposition of tetraethyl orthosilicate at 900°C) and elemental silicon (by hydrogen reduction of trichlorosilane at 900°C). Although these coatings imparted increased oxidation resistance to the fibers, there was no significant change in

composite shear properties. Similar negative results have been obtained
with electrochemically deposited metal coatings [42]. High-modulus acrylic-
based carbon fibers have been coated with pyrolytic carbon (PC) by pyrolyzing
either acetylene or methane at 1000°C [63]. The wettability of carbon films
so deposited on quartz was measured and found to have a critical surface
tension of 45 ± 2 dynes/cm. Such coatings should be wetted by epoxies
($\gamma \simeq 48$ dynes/cm), producing a better bond and hence improving the composite
shear properties. An improvement was observed with PC-coated Modmor I
fibers (Table 19), but a slight detrimental effect was observed with surface-
treated Modmor II fibers, the latter probably being due to an adverse reaction between the treated surface and the coating.

TABLE 19

Properties of Epoxy-Carbon Fiber Composites

Sample No.	Flexural modulus (10^6 psi)	Flexural strength (10^3 psi)	Interlaminar shear strength (10^3 psi)
Modmor I Fibers Coated with Pyrolytic Carbon			
1	28.1	150	5.1
2	28.5	128	6.0
3	28.5	121	6.2
Average	28.4	133	5.8
Control[a]	28.0	110	3-4
Modmor II (S)[b] Fibers Coated with Pyrolytic Carbon			
1	18.5	306	11.8
2	19.1	324	12.8
3	19.1	317	11.9
4	19.5	295	12.1
Average	19.0	310	12.2
Control[a]	19.0 ± 1	310 ± 10	13.5 ± 5

[a]Uncoated
[b]These fibers had been given some type of proprietary surface treatment by
the supplier.

3. Whiskerizing

A new technique developed at the U.S. Naval Ordnance Laboratory shows considerable promise of improving the shear properties of carbon-fiber composites. This process, known as "whiskerizing," involves the growth of single crystals of β-silicon carbide on the fiber surface perpendicular to the fiber axis. Details of the process have not been described, although the technique involves vapor deposition of silicon carbide from a hydrogen stream at 1100 to 1700°C [57]. The whiskers grow out from the individual fibers in a radial arrangement, as shown in Fig. 40.

Dramatic increases in the interlaminar shear strengths of composites made from whiskerized carbon fibers have been reported [57, 72, 73]. The effect of this treatment on the shear strengths of various fibers is shown in Fig. 41. Whiskerizing apparently leads to a threefold increase in shear strength for Thornel 25 composites and a fivefold improvement for RAE (PAN-based) fibers. The improvement in mechanical properties is probably the result of the three-dimensional network of silicon carbide whiskers, the strong bonding of whiskers to the graphitic substrate, and the enormously increased interfacial area, rather than any increase in bonding between fiber and resin matrix. In this respect the nature of the effect is fundamentally different from that produced by oxidative chemical treatments.

VI. FUNCTIONAL GROUPS ON GRAPHITIC SURFACES

The presence of polar surface groups is frequently invoked to explain the surface properties of carbon fibers that have been oxidized according to the methods described in the preceding section. It therefore seems appropriate to review briefly the variety of chemical groups that have been introduced by oxidative treatment of carbon surfaces and the methods that have been used to characterize them. Several excellent reviews of this large subject have appeared recently [75, 76], and only a brief summary of relevance to carbon fibers will be attempted here.

It is well known that the surface properties of carbonaceous or graphitic materials depend to a large extent on the presence or absence of "surface oxides," which can be introduced or removed by appropriate pretreatment. Functional groups formed by oxidation make the carbon hydrophilic, presumably by providing sites for the chemisorption of water molecules. It is this changed character of the carbon surface, together with the increase in

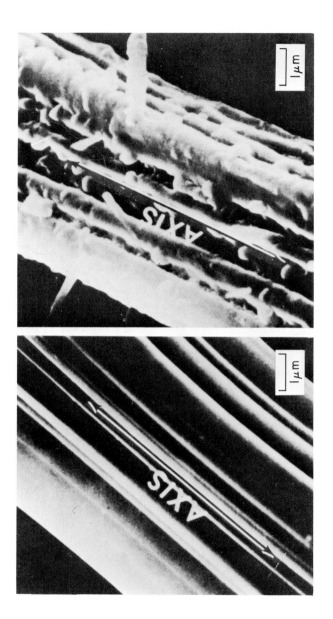

FIG. 40.   Scanning electron micrographs Thornel carbon fiber (left) and silicon carbide whiskers growing on the same type of fiber (right).   Reprinted by permission from Ref. [57].

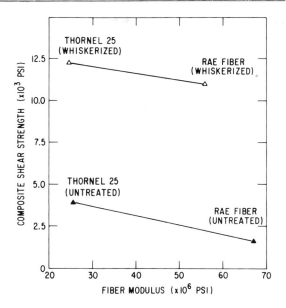

FIG. 41. Effect of whiskerizing on carbon fiber-epoxy composite shear strength. From Ref. [74].

surface area accompanying the oxidation process, that is often held respons-ible for the observed increased composite shear strengths of treated carbon fibers.

Several types of polar groups have been positively identified on oxidized carbon surfaces. The most common species are carboxyl, R-COOH, phenolic hydroxyl, R-OH, and carbonyl, $R \gtrless C = O$, R being the graphitic substrate in each case. Other possible groups, the identification of which is less firmly established, include lactones, possibly a condensation product of adjacent carbonyl and carboxyl groups, (1); quinoid structures, (2) [77] from the oxidation of adjacent phenolic hydroxyls; and hydroperoxides, R-C-O-H [78].

(1)   (2)

The experimental methods for identifying and determining the more important of these groups are discussed in this section.

A. Determination of Carboxyl

The strongly acid carboxyl groups can be titrated by means of sodium bicarbonate, which does not react with phenolic hydroxyl. The most usual titration method, which has been frequently employed with carbon fibers [42], was developed by Donnet et al. [79]. The fibers are refluxed with an excess of calcium acetate in a water slurry for 24 h. Acetic acid formed by the reaction

$$2R\text{-}COOH + (CH_3COO)_2Ca \longrightarrow (RCOO)_2\ Ca + 2CH_3COOH$$

is then titrated with N/50 potassium hdyroxide.

Other methods for estimation of surface carboxyl include methylation with diazomethane

$$R\text{-}COOH + CH_2N_2 \longrightarrow R\text{-}COOCH_3 + N_2$$

The methyl esters formed can then be hydrolyzed with dilute hydrochloric acid. Also used is reaction with thionyl chloride to give an acid chloride

$$R\text{-}COOH + SOCl_2 \longrightarrow R\text{-}COCl + SO_2 + HCl$$

which can be characterized by further reaction with dimethylaniline to give an anilide [80].

Carboxyl groups on carbon surfaces decompose to carbon dioxide at elevated temperatures, the decomposition being essentially complete at $900^{\circ}C$. It has been suggested by Puri and Bansal [81] that surface "carboxyl" is nothing more than strongly chemisorbed carbon dioxide. This interpretation offers no explanation for the characteristic chemical and hydrophilic properties of oxidized carbon surfaces.

B. Determination of Hydroxyl

Surface phenolic hydroxyls appear to be neutralized by sodium hydroxide, but not by sodium carbonate, so that the difference between the NaOH and Na_2CO_3 titration values gives a measure of phenolic hydroxyl concentration [75]. These groups can also be methylated with diazomethane

$$R\text{-}OH + CH_2N_2 \longrightarrow R\text{-}OCH_3 + N_2$$

the resultant methoxy groups being much more resistant to acid hydrolysis than the methyl esters derived from surface carboxyls. These methods have been used in the estimation of surface hydroxyl on carbon fibers [57].

C. Determination of Carbonyl

There is still considerable disagreement concerning the nature of carbonyl groups on oxidized carbon surfaces. The presence of carbonyl oxygen on oxidized graphite has been detected by infrared spectroscopy [82], but the surface coverage is generally low. Carbon surfaces containing carbonyl groups often show quite varied chemical porperties, such as those attributed to acid anhydrides, lactones, or quinones.

Carbonyl surface concentrations have been determined by a method developed by Studebaker [83], involving reaction with diazomethane to give a cyclic pyrazoline:

$$\text{(quinone)} + CH_2N_2 \longrightarrow \text{(pyrazoline product)}$$

Carbonyl groups on carbon have also been estimated by means of a hydrogen titration, using a platnium hydrogenation catalyst

$$\text{(quinone)} + H_2 \xrightarrow{Pt} \text{(hydroquinone)}$$

or by means of the reaction with sodium borohydride

$$4\ \text{(quinone)} + NaBH_4 + 2H_2O \longrightarrow 4\ \text{(hydroquinone)} + NaBO_2$$

None of these methods has, however, as yet been used with carbon fibers.

D. Distribution of Functional Groups on Oxidized Carbon Surfaces

The determination of carboxyl or total acid groups by base titration is often the only procedure used to estimate the surface functionality of carbon fibers. However, sufficient information is available on the behavior of other carbon and graphite surfaces to assess the relative proportions of various groups produced by different oxidizing agents. Further estimates can also be made by analysis of the gases desorbed from fibers at elevated temperatures.

1. Oxidation by Gaseous Oxygen and Air

The functional groups produced by the low-temperature oxidation of a natural graphite powder were studied by Kiselev et al. [84]. Natural graphite crystals were ground in an argon atmosphere to a surface area of 380 m^2/g. Oxygen and water vapor were then admitted at room temperature. The resulting surface functional groups are listed in Table 20. In this case carboxyl groups were estimated by titration with sodium carbonate, hydroxyl from the difference in titration values between NaOH and Na_2CO_3, hydroperoxide by iodometric titration in aqueous isopropanol, and carbonyl by reaction with p-bromophenyl magnesium bromide. The oxygen contained in these groups accounted for only 29% of the total oxygen content of the oxidized graphite, and it is possible that other groups and radicals were also present. On outgassing of the sample above 100°C, oxygen, carbon monoxide, and carbon dioxide were evolved, as shown in Fig. 42 [85].

TABLE 20

Functional Groups on Oxidized
Cold-Milled Graphite[a]

Group	Content (meq/m^2)
$\overset{\displaystyle O}{\overset{\displaystyle \|}{-COH}}$	0.48
$\geq$C-OH	0.16
$\geq$C-OOH	0.04
$\geq$C=O	0.07

[a]Data from Ref. [84].

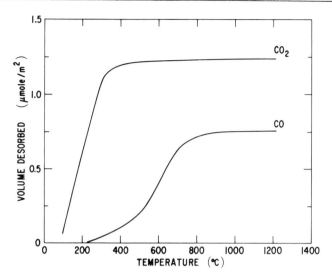

FIG. 42. Desorption of gases from natural graphite powder on heating. After Kiselev et al. [85].

The distribution of desorbed products from oxidized graphitic materials has been studied in detail by Lang and associates [86]. For a purified nuclear-grade graphite and a graphitized carbon black (Spheron 6) after oxidation in air at $620°C$, the composition of the desorbed gas at various temperatures was as shown in Figs. 43 and 44. In each case the maximum evolution of gas occurred at 650 to $700°C$, and the major desorbed product was carbon monoxide. It appears that under these conditions of oxidation the major surface functional group formed is not carboxyl. In each case the oxidation was accompanied by a considerable increase in surface area. This is of interest in view of the finding of Hennig [87] that functional groups are located mostly on the edge atoms of the graphite planes. Thus studies of graphite single crystals showed that below $800°C$ the oxidation rate for carbon atoms at edge surfaces was at least 20 times higher than that for those in the basal planes. As oxidation would be expected to cause pitting on the basal planes and the appearance of fresh prismatic surfaces [61] for the attachment of functional groups, the simultaneous increase in surface area and functionality is not unexpected.

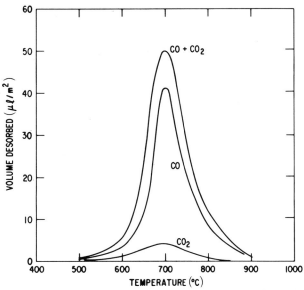

FIG. 43. Desorption of gases from nuclear graphite after oxidation in air at $620^\circ C$. After Lang et al. [86].

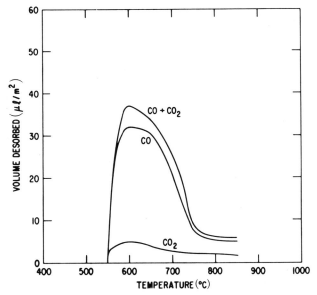

FIG. 44. Desoprtion of gases from graphitized carbon black (Spheron 6) after oxidation in air at $620^\circ C$. After Lang et al. [86].

The discrepancy in the relative proportions of desorbed carbon monoxide and carbon dioxide shown in Figs. 42 and 43 suggests that the distribution of functional surface groups is very sensitive to the nature of the graphite surface and the conditions of oxidation.

2. Oxidation by Nitric Acid

Nitric acid oxidation of graphite appears to result in a greater proportion of surface carboxyl than in the case of oxidation with gaseous oxygen or air. Thus treatments of various carbon blacks with boiling 30% or 40% nitric acid caused a general diminution in the diameter of the particles and carboxyl-hydroxyl ratios of 7 to 18 in the oxidized products [79, 88]. The distribution of desorbed products from nitric acid-oxidized Spheron 6 graphitized carbon is shown in Fig. 45 [86], the proportion of carbon dioxide being as much as 50% in this case. The total amount of carbon dioxide desorbed from this sample agreed very well with the surface carboxyl concentration determined by titration.

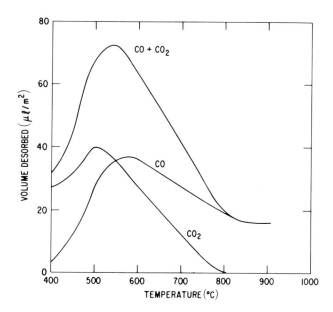

FIG. 45. Desorption of gases from graphitized carbon black (Spheron 6) after oxidation with nitric acid. After Lang et al. [86].

Although both air and nitric acid oxidation of carbon fibers are known to result in increased amounts of total acid groups [42, 57], the relative proportions of the various groups have not been investigated. In addition the simultaneous increase in surface area of the fibers may result in a net decrease in functionality per unit area [58].

The temperature dependence of gas evolution from untreated and treated Modmor I fibers is shown in Fig. 46. The total gas evolved on heating the samples to $900^{\circ}C$ is shown in Table 21.

TABLE 21

Gas Evolved from Treated Modmor I Fibers[a]

Treatment	Gas evolved (ml/g, STP)
None	0.46
Supplier's	1.23
60% HNO_3 72 h at $118^{\circ}C$	1.16

[a] From Ref. [63].

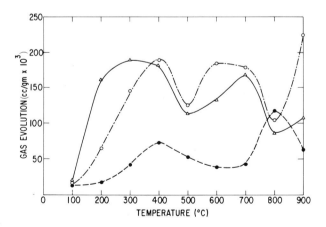

FIG. 46. Temperature dependence of gas evolution from Modmor I carbon fibers. Key: ----, untreated; —·—, supplier's surface treatment; ———, after 72-h treatment with 60% HNO_3 at $118^{\circ}C$.

Below $500^\circ C$, carbon dioxide was the major constituent probably coming from decomposition of surface carboxyl groups. Above $500^\circ C$, carbon monoxide was the major species evolved, with hydrogen making a significant contribution (20-25%) above $800^\circ C$.

3. Oxidation by Other Aqueous Oxidants

Although comparable data are not available for carbon fibers, Donnet and associates [79] have estimated the proportions of carboxyl and phenolic hydroxyl produced on the surface of carbon black by treatment with various aqueous oxidizing agents. The results are summarized in Table 22.

TABLE 22

Functional Groups Formed on Carbon Black[a]

Reagent	Temperature	COOH (g/100 g)	OH (g/100 g)	COOH/OH
$KMnO_4/4$ N H_2SO_4	Ambient	5.7	0.6	9.5
$NaClO_2/4$ N H_2SO_4	Ambient	1.3	0.3	4.3
$K_2Cr_2O_7/$ 4 N H_2SO_4	Ambient	1.3	0.6	2.2
NaOCl	$0^\circ C$	1.3	0.4	3.2
N/5 $Na_2S_2O_8/4$ N H_2SO_4, Ag^+	$80^\circ C$	1.6	0.7	2.3

[a]Data from Ref. [79]. Philblack O carbon black.

As many of these oxidizing agents have been used in the surface treatment of carbon fibers, it is tempting to assume that the resulting changes in composite shear properties are related to the concentration and chemical nature of the functional surface groups which bond with the resin matrix. However, because of the effects of increased surface area and the observed fact that composite shear properties are not always significantly reduced after heat-treating the fibers at high temperatures [63], this conclusion is unwarranted at the present time.

VII. ADSORPTION PROPERTIES AND REACTIVITY OF CARBON FIBERS

A. Adsorption Measurements

Apart from determinations of nitrogen and krypton adsorption, few measurements of the adsorption properties of carbon fibers have been made. This is surprising in view of the large amount of useful information concerning the shapes and sizes of micropores and the energetics of surfaces that can be obtained from adsorption isotherms.

The physical adsorption of several hydrocarbon vapors on Saran carbon fibers has recently been studied by Adams, Boucher, and Everett [43]. The fibers were prepared from Saran precursor (a copolymer of polyvinyl and polyvinylidene chlorides) by carbonization at various temperatures in the range 158 to $179^{\circ}C$, followed by heat treatment at $760^{\circ}C$ in nitrogen for 12 h. The fibers were not graphitized and were essentially amorphous. Adsorption isotherms of benzene, n-hexane, cyclohexane, and 2,2-dimethylbutane at $25^{\circ}C$ on a typical Saran carbon fiber are shown in Fig. 47. Although adsorption

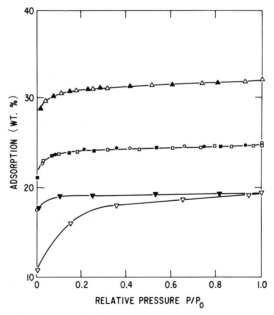

FIG. 47. Adsorption of hydrocarbon vapors on Saran carbon fibers (F158) at $25^{\circ}C$. Key: $\triangle$, benzene; $\square$, h-hexane; o, cyclohexane; ∇, 2,2-dimethylbutane. Filled symbols denote desorption. After Adams et al. [43].

of benzene, n-hexane, and cyclohexane was quite rapid, adsorption of 2,2-
dimethylbutane took over 100 h to reach equilibrium at each adsorption
point, and a large hysteresis effect was observed. Adsorption of isooctane
by the fibers was so slow that equilibrium isotherms could not be obtained.
The volumes of these adsorbates taken up at saturation by three different
fiber samples are listed in Table 23. The mean pore volumes of the Saran
carbon fibers were in the range 0.3 to 0.4 ml/g, which may be compared
with the much lower value of 0.003 ml/g found with Thornel 40 (see Table 3).
Also the density of the Saran-based fibers was only 1.3 g/ml, compared with
about 1.6 g/ml for rayon-based fibers (Table 4). It appears that the porosity
of the former is confined to micropores in the 10- to 20- Å range [43].

A study has recently been made [89] of the surface reactivity of a number of commercial carbon fibers toward water and several organic vapors
(toluene, pyridine, n-decane, isobutyric acid, n-octylamine, and aniline).
The fibers were used as adsorbent columns in a gas chromatograph, and the
adsorption affinities were determined from the retention times for the various fibers. Heats of adsorption derived from the temperature dependence of
the retention times did not vary significantly for various surface treatments
of the fibers. However, calculated adsorption coefficients indicated that
oxidation of the fibers with nitric acid or air enhanced the adsorption of such
polar adsorbates as water, pyridine, and aniline. More recently Belinski
et al. [90] measured heats of adsorption of carbon tetrachloride, methyl

TABLE 23

Adsorption Volumes of Saran-Based Carbon Fibers[a]

Adsorbate	T (^{o}C)	Adsorption volume (ml/g)		
		F 158[b]	F 167[b]	F 178[b]
Nitrogen	-196	0.407	0.398	0.400
Benzene	25	0.369	0.372	0.370
n-Hexane	25	0.379	0.380	0.382
Cyclohexane	25	0.323	0.317	0.327

[a]Data from Ref. [42].

[b]Numbers refer to carbonization temperature.

ethyl ketone, dioxane, benzene, and thiophene on Courtaulds HT carbon fiber by a chromatographic method. The results were very similar to those reported previously for graphitized carbon black.

B. Intercalation Compounds

Carbon fibers, in common with natural graphite, have been shown to form intercalation compounds with alkali metals, such as potassium and cesium [91, 92]. Ruland and associates [91] used an automatic spring balance and showed that, although at 125°C the absorption of potassium vapor by Thornel fibers was extremely slow, at 140°C absorption began at a potassium pressure between 10^{-5} and 10^{-4} torr, and increased up to a K/C ratio of 0.15. At 200°C physical adsorption was negligible, and the composition of the saturated fiber corresponded to C_8K. X-Ray studies of this product showed alternating carbon and potassium layers, the latter having a two-dimensional hexagonal-close-packed structure, as in the corresponding compound obtained from graphite. Other phases, such as $C_{16}K$ and $C_{24}K$, were also identified.

Although the presence of potassium had a vary marked effect in increasing the electrical conductivities of the carbon fibers, neither the preferred orientation of the carbon layers nor the pore structure was appreciably affected by the intercalation of potassium or cesium. These results indicate that there are essentially no crosslinks perpendicular to the graphitic layers, which can move easily relative to each other perpendicular to the fiber axis [2].

C. Reaction with Oxygen

Although mild oxidation of carbon fibers in gaseous oxygen can be carried out at elevated temperatures without serious damage to the fiber structure, at temperatures much in excess of 600°C carbon fibers, in common with other forms of carbon and graphite, will ignite and burn spontaneously in air or oxygen.

Thermograms for Modmor I and II fibers heated at a linear rate of 300°C/h are shown in Figs. 48 and 49 for flowing oxygen and air, respectively [63]. Modmor II showed substantially lower ignition temperatures in both gases than the more highly graphitized Modmor I, probably as a result of the large proportion of more reactive amorphous carbon in the former fiber. The oxidation characteristics of Thornel 25 were very similar to those of Modmor I.

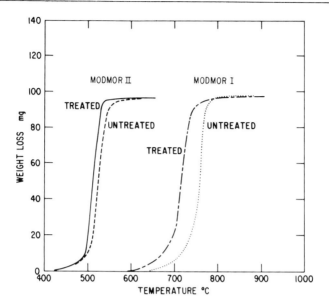

FIG. 48. Thermograms for the oxidation of Modmor carbon fibers in oxygen. Heating rate $300^{\circ}C/h$, oxygen flow rate 250 ml/min.

D. Reaction with Hydrogen

Natural graphite crystals heated in flowing hydrogen gas at $720^{\circ}C$ show no observable changes in surface structure after 8-h exposure [93]. We have found, however, that PAN-based Modmor II fibers show a gradual loss in weight on heating in flowing hydrogen to temperatures exceeding $700^{\circ}C$, the mean rate of weight loss at $1000^{\circ}C$ being 2.3 %/h, as determined by isothermal thermogravimetry [63]. The change in appearance of the fibers after this treatment is shown in Fig. 50. In addition to a uniform decrease in fiber diameter, a general smoothing of the fiber surface is apparent.

VIII. CONTACT ANGLE AND WETTING PROPERTIES OF CARBON FIBERS

The contact angle is a common and useful measure of wettability. It was Bernett and Zisman [94] who observed that a linear relationship existed between the cosine of the contact angle and the surface tension against air

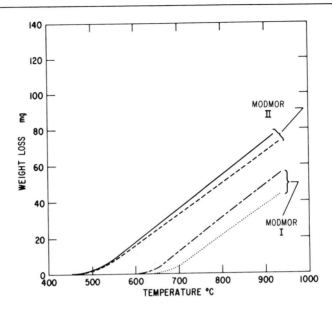

FIG. 49. Thermograms for the oxidation of Modmor carbon fibers in air. Heating rate = 300°C/h, air flow rate - 250 ml/min.

for a variety of liquids, and the intercept at cos $\theta = 1$ is chosen as the critical surface tension γ_c of the solid; that is, liquids whose surface tension is less than the critical surface tension of the solid will spread spontaneously at a rate determined by their viscosity.

The relationship between wettability (adhesion) and contact angle is well documented for large flat specimens [95-97], but the method is difficult to apply experimentally to small-diameter fibers. In this area the technique has been useful for investigating the effectiveness of surface treatments of glass fibers, and the agreement with their macroscopic flat analogs is fairly good [98, 99].

As described in Section II. F, Bobka and Lowell [34] measured the rise of liquid as a function of distance from the axis of the filament. By fitting the data with a fourth-order Runge-Kutta approximation, they were able to determine the angle of intercept of resin and fiber. Their results are shown in Table 24.

FIG. 50. Surface topography of Modmor II carbon fibers before (top) and after (bottom) reduction with H_2 at $1000^\circ C$ for 2 h. x 6000

TABLE 24

Contact Angle between Liquids and Thornel 25 Carbon Fibers

Fiber treatment	Liquid[a]	Contact angle (degrees)
None	H_2O	36
	ERLA-0400	12 ± 7
	ERL-2774	32 ± 8
15 min in O_2 at 500° + thermal desorption	H_2O	38 ± 8
	ERLA-0400	4.3 ± 0.6
	ERL-2774	5.4 ± 0.7

[a]ERLA-0400 and ERL-2774 are epoxy resins.

The data in Table 24 indicate that surface oxidation improves the wettability of the fibers especially with respect to epoxy resins. Bobka and Lowell [34] also measured the amount of liquid absorbed by a yarn bundle when immersed for a definite period of time. Using liquids of various physical properties, they concluded that viscosity controls the wicking rates.

Goan and Prosen [57] have used Zisman's procedure to determine the critical surface tension γ_c of wetting. They obtained a value of 46 dynes/cm for carbon fibers. Since most epoxy resins have a surface tension ranging from 44 to 48 dynes/cm, the fibers should be adequately wetted by epoxy resins.

We used Zisman's procedure to measure the wettability of carbon films pyrolytically deposited on quartz [63]. The critical surface tension of the film was 45 ± 2 dynes/cm, which is similar to the value reported by Goan and Prosen [57]. Commercially available high-modulus carbon fibers were coated with pyrolytic carbon, incorporated into an epoxy matrix, and tested in flexure. The pyrolytic carbon coating improved the interlaminar shear strength by approximately 20%.

Bobka and Lowell [34] used a surface-velocity technique (Section II. F. 2.b) to measure the effect of surface treatments on the wicking rates of graphitic yarn. The results are shown in Fig. 51.

Recently there has been some work on the wetting of carbon by copper alloys in connection with the metal impregnation of carbon-fiber composites

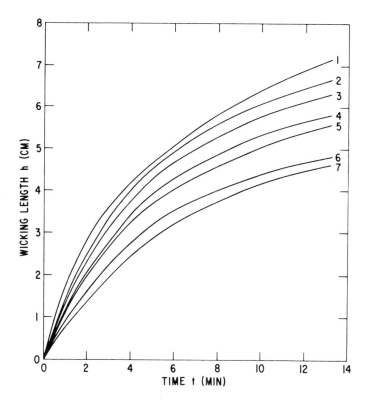

FIG. 51. Effect of surface treatments on the wicking rates of graphitic yarn. Curve 1, 15-min treatment with $Na_2Cr_2O_7$-H_2SO_4 at 100°C; curve 2, 5-min treatment with $Na_2Cr_2O_7$-H_2SO_4 at 100°C; curve 3, 15-min treatment with $KMnO_4$-H_2SO_4 at 60°C; curve 4, 5-min treatment with $KMnO_4$-H_2SO_4 at 60°C; curve 5, 5-min treatment with alcoholic NaOH at 50°C; curve 6, untreated yarn; curve 7, yarn boiled in water for 15 min at 100°C. After Bobka and Lowell [34].

[100]. Pure copper is inert and nonwetting, but two additions caused the copper to wet: chromium on both HX30 graphite and vitreous carbon and vanadium on the latter. In this case wettability was determined by measuring sessile-drop contact angles.

IX. GENERAL CONCLUSIONS

Strong, highly oriented carbon fibers produced by heat treatment of organic polymer fibers at elevated temperatures possess a structure that is unique among the known forms of carbon. The remarkable mechanical properties are a direct consequence of the unusual arrangement of the graphitic subunits into elongated ribbonlike fibrils that run for long distances parallel to the fiber axis. The surface properties of the fibers show many similarities to those of graphite and carbon black, and can be altered fundamentally by various chemical reagents. Although changes in surface area and surface functionality have been invoked to explain the increase in composite shear strength after oxidative treatments, the importance, if any, of these factors is not known in every case. At least with PAN-based fibers the lack of correlation between composite shear strengths and surface area and functionality changes resulting from oxidative treatments suggests that chemical treatments may merely remove defects in the fiber structure which act as local points of weakness. Detailed studies of simultaneous changes in surface area, pore-size distribution, surface functionality, and composite shear properties after oxidative treatments under carefully controlled conditions are needed to establish unequivocally the effects of surface treatments on fibers of different types. Much fundamental research remains to be done on the surface energetics of carbon fibers and the relation between structure and mechanical properties, and there is every reason to believe that such information will be forthcoming as the applications of carbon fibers in structural materials accelerates during the next few years.

ACKNOWLEDGMENTS

The authors are most grateful to Drs. B. W. Roberts and V. A. Phillips for kindly providing several electron micrographs and to Drs. J. R. McLoughlin, E. A. Boucher, and D. A. Scola for making data available prior to publication. The scanning electron micrographs were prepared by Dr. E. Lifshin.

Thanks are also due to Pergamon Press, Technomic Publishing Company, Interscience Publishers, the Institute of Physics, the Editors of Nature, and the American Society for Testing and Materials for permission to reproduce certain figures.

REFERENCES

[1]   R. Bacon, High Temperature Resistant Fibers from Organic Polymers, Interscience, New York, 1969, p. 213.

[2]   A. Fourdeux, C. Hérinckx, R. Perret, and W. Ruland, Compt. Rend., 269, 1597 (1969).

[3]   S. Ergun, in Chemistry and Physics of Carbon (P. L. Walker, Jr., ed.), Vol. 3, Dekker, New York, 1968, p. 211.

[4]   W. Ruland, in Chemistry and Physics of Carbon (P. L. Walker, Jr., ed.), Vol. 4, Dekker, New York, 1968, p. 1.

[5]   W. T. Brydges, D. V. Badami, J. C. Joiner, and G. A. Jones, in High Temperature Resistant Fibers from Organic Polymers, Interscience, New York, 1969, p. 255.

[6]   D. J. Johnson and C. N. Tyson, Brit. J. Appl. Phys. (J. Phys. D.), 2, 787 (1969).

[7]   R. E. Franklin, Acta Cryst., 3, 107 (1950); B. E. Warren, Phys. Rev., 59, 693 (1941).

[8]   G. E. Bacon, J. Appl. Chem., 6, 477 (1956).

[9]   D. H. Logsdail, in High Temperature Resistant Fibers from Organic Polymers, Interscience, New York, 1969, p. 245.

[10] M. Yamamoto and S. Yamada, in High Temperature Resistant Fibers from Organic Polymers, Interscience, New York, 1969, p. 263.

[11] W. Ruland, J. Appl. Phys., 38, 3585 (1967).

[12] P. Debye and A. Bueche, J. Appl. Phys., 20, 518 (1949).

[13] G. Porod, Kolloid-Z., 124, 83 (1951); 125, 51, 109 (1952).

[14] D. Tchoubar-Vallat and J. Méring, Compt. Rend., 261, 3096, 3361 (1965).

[15] M. M. Tang and R. Bacon, Carbon, 2, 211, 221 (1964).

[16] D. V. Badami, C. Campbell, A. D. Day, and M. J. Lindsey, in Second Conf. Ind. Carbon and Graphite, 1965, Society of Chemical Industry, London, 1966.

[17] P. D. Frayer and J. B. Lando, J. Coll. Interface Sci., 31, 145 (1969).

[18] A. Fourdeux, R. Perret, and W. Ruland, 9th Conf. Carbon, Boston, 1969, Paper SS-23.

[19] W. Watt and W. Johnson, High Temperature Resistant Fibers from Organic Polymers, Interscience, New York, 1969, p. 215.

[20] W. Johnson and W. Watt, Nature, 215, 384 (1967).

[21] R. B. Bolon, J. A. Hugo, V. A. Phillips, R. R. Russell, and B. W. Roberts, unpublished work.

[22] D. V. Badami, J. C. Joiner, and G. A. Jones, Nature, 215, 386 (1967).

[23] B. L. Butler and R. J. Diefendorf, 9th Conf. Carbon, Boston, 1969, Paper SS-25.

[24] R. D. Heidenreich, W. M. Hess, and L. L. Ban, J. Appl. Cryst., 1, 1 (1968).

[25] P. E. Trent and R. K. Bennett, 9th Conf. Carbon, Boston, 1969, Paper SS-24.

[26] B. Harris, P. W. R. Beaumont, and A. Rosen, J. Mater. Sci., 4, 432 (1969).

[27] J. W. Johnson, in High Temperature Resistant Fibers from Organic Polymers, Interscience, New York, 1969, p. 229.

[28] P. W. Jackson, Metals Eng. Quart., 9(3), 22 (1969).

[29] D. Robson, F. Y. I. Assabghy, D. J. E. Ingram, and P. G. Rose, Nature, 221, 51 (1969).

[30] G. Wagoner, Phys. Rev., 118, 647 (1960).

[31] M. Barber, P. Swift, E. L. Evans, and J. M. Thomas, Nature, 227, 1131 (1970).

[32] G. L. Connell, Nature, 230, 377 (1971).

[33] F. Tuinstra and J. L. Koenig, J. Comp. Mater., 4, 492 (1970).

[34] R. J. Bobka and L. P. Lowell, Integrated Research on Carbon Composite Materials, AFML-TR-66-310, Part I, Air Force Materials Laboratory, Oct. 1966, pp. 145-152.

[35] E. W. Washburn, Phys. Rev., 17(3), 273 (1921).

[36] R. L. Peek and D. A. McLean, Ind. Eng. Chem., 6, 85 (1934).

[37] R. Didchenko, Carbon and Graphite Surface Properties Relevant to Fiber Reinforced Composites, AFML-TR-68-45, Air Force Materials Laboratory, Feb. 1968.

[38] J. C. Goan and S. P. Prosen, 71st ASTM Meeting, San Francisco, 1968.

[39] V. J. Mimeault and D. W. McKee, Nature, 224, 793 (1969).

[40] A. J. Rosenberg, J. Amer. Chem. Soc., 78, 2929 (1956).

[41] J. W. Herrick, 23rd Ann. Tech. Conf. SPI Reinforced Plastics/ Composites Division, Section 16-A, February 1968.

[42] J. W. Herrick, P. E. Gruber, Jr., and F. T. Mansur, Surface Treatments for Fibrous Carbon Reinforcements, AFML-TR-66-178, Part I, Air Force Materials Laboratory, July 1966.

[43] L. B. Adams, E. A. Boucher, and D. H. Everett, Ninth Conference on Carbon, Boston, 1969, Paper SP-11.

[44] D. H. Everett and W. I. Whitton, Proc. Roy. Soc. (London), A230, 91 (1955).

[45] W. Ruland, J. Polymer Sci., No. 28, Part C, 143 (1969).

[46] D. L. Schmidt and H. T. Hawkins, Filamentous Carbon and Graphite, AFML-TR-65-160, Air Force Materials Laboratory, August 1965.

[47] E. Barrett, L. Joyner, and P. Halenda, J. Amer. Chem. Soc., 73, 373 (1951).

[48] B. F. Roberts, J. Coll. Interface Sci., 23, 266 (1967).

[49] P. W. Jackson and J. R. Marjoram, Nature, 218, 83 (1968).

[50] P. W. Jackson and J. R. Marjoram, J. Mater. Sci., 5, 9 (1970).

[51] R. A. Coyle, L. M. Gillin, and B. J. Wicks, Nature, 226, 257 (1970).

[52] W. Ruland, 9th Conf. Carbon, Boston, 1969, Paper MP-24.

[53] R. Perret and W. Ruland, 9th Conf. Carbon, Boston, 1969, Paper SS-22.

[54] S. Ergun, 9th Conf. Carbon, Boston, 1969, Paper SS-21.

[55] J. A. Hugo, V. A. Phillips, and B. W. Roberts, Nature, 226, 144 (1970).

[56] A. A. Pallozzi, Soc. Plastics Eng. J., 80, February 1966.

[57] J. C. Goan and S. P. Prosen, in Interfaces in Composites, ASTM Special Tech. Publ. No. 452, p. 3.

[58] D. A. Scola and C. S. Brooks, 25th Soc. Plastics Ind. Meeting, Washington, D.C., February 1970.

[59] J. Butcher and D. M. Grove, in Proc. 5th Conf. Carbon, Vol. 1, Pergamon Press, Oxford, 1962, p. 205.

[60] S. P. Prosen, J. V. Duffy, P. W. Erickson, and M. A. Kinna, 21st Soc. Plastics Ind. Ann. Tech. Conf., 1966, Section 8-D.

[61] J. M. Thomas, in Chemistry and Physics of Carbon (P. L. Walker, Jr., ed.), Vol. 1, Dekker, New York, 1965, p. 121.

[62] H. Wells and W. J. Colclough, Ger. Offen. 1, 817, 581 October 1969; Chem. Abstr., 71, 125895t (1969).

[63] V. J. Mimeault and D. W. McKee, unpublished results.

[64] R. C. Novak, in Composite Materials: Testing and Design, ASTM Special Tech. Publ. No. 460, 1969, p. 540.

[65] N. J. Wadsworth and W. Watt, U.S. Pat. 3,476,703 (1969).

[66] R. S. Sach and J. Bromley, Ger. Offen., 1,817,578, August 1969; Chem. Abstr., 71, 103026h (1969).

[67] T. Dohi, M. Asano, and S. Magari, Tanso, 57, 192 (1969); Chem. Abstr., 72, 4764 (1970).

[68] E. Papirer, J. B. Donnet, and A. Schultz, Carbon, 5, 113 (1967).

[69] D. W. McKee, Carbon, 8, 623 (1970).

[70] D. W. McKee, Carbon, 8, 131 (1970).

[71] J. W. Herrick, 12th Natl. Symp. Soc. Aerosp. Mater. Process Eng., Anaheim, Calif., October 1967, Paper AC-8.

[72] R. G. Shaver, A.I. Chem. E. Materials Conf., Philadelphia, April 1968.

[73] R. A. Simon and S. P. Prosen, 23rd Ann. Tech. Conf., Soc. Plastics Ind., Reinforced Plastics Div., Section 16-B, February 1968.

[74] C. Z. Carroll-Porczynski, Advanced Materials, 2nd ed., Chemical Publishing, New York, 1969.

[75] H. P. Boehm, Adv. Catalysis, 16, 179 (1966); Angew. Chem. Intern. Ed., 5, 533 (1966).

[76] J. B. Donnet, Carbon, 6, 161 (1968).

[77] V. A. Garten and D. E. Weiss, Aust. J. Chem., 8, 68 (1955).

[78] Y. A. Zarif'yants, V. F. Kiselev, N. Lezhnev, I. Novikova, and G. G. Fedorov, Dokl. Akad. Nauk SSSR, 143, 1358 (1962).

[79] J. B. Donnet, F. Hueber, C. Reitzer, J. Oddoux, and G. Reiss, Bull. Soc. Chim. France, 1727 (1962).

[80] J. B. Donnet, French Pat. 1,164,786 (1958).

[81] B. R. Puri and R. C. Bansal, Carbon, 1, 457 (1964).

[82] J. S. Mattson and H. B. Mark, Jr., J. Coll. Interface Sci., 31, 131 (1969).

[83] M. L. Studebaker, Rubber Age, 77, 69 (1955).

[84] Y. A. Zarif'yants, V. F. Kiselev, N. Lezhnev, and G. G. Fedorov, Dokl. Akad. Nauk SSR, 143, 1358 (1962).

[85] Y. A. Zarif'yanz, V. F. Kiselev, N. Lezhnev, and O. Nikitina, Carbon, 5, 127 (1967).

[86] F. M. Lang, M. De Noblet, J. B. Donnet, J. Lahaye, and E. Papirer, Carbon, 5, 47 (1967).

[87] G. Hennig, in Proc. 5th Conf. Carbon, Vol. 1, Pergamon Press, Oxford, 1962, p. 143.

[88] J. B. Donnet, F. Hueber, N. Perol, and J. Jaeger, J. Chim. Phys., 60, 426 (1963).

[89] C. S. Brooks and D. A. Scola, J. Coll. Interface Sci., 32, 561 (1970).

[90] C. Belinski, C. Diot, and F. X. L. Keraly, Compt. Rend., C271, 1025 (1970).

[91] C. Hérinckx, R. Perret, and W. Ruland, Nature, 220, 63 (1968).

[92] C. Hérinckx, R. Perret, and W. Ruland, 9th Conf. Carbon, Boston, 1969, Paper RS-25.

[93] B. McCarroll and D. W. McKee, Nature, 225, 723 (1970).

[94] M. K. Bernett and W. A. Zisman, J. Phys. Chem., 63, 1241, (1959); W. A. Zisman, Contact Angle, Wettability, and Adhesion, Advances in Chemistry Series No. 43, American Chemical Society, Washington, D.C., 1964.

[95] H. Schonborn and F. W. Ryan, J. Adhesion, 1, 43 (1969).

[96] H. Schonborn, H. L. Frisch, and T. K. Kwei, J. Appl. Phys., 37, 4967 (1966).

[97] R. E. Johnson, Jr., and R. D. Dettre, J. Adhesion, 2, 3 (1970).

[98] W. D. Bascom, Symp. on the Interface in Composites, 42nd Natl. Colloid Symp., Chicago, June 1968.

[99] P. E. Throckmorton and M. F. Brown, Proc. SPI Reinforced Plastics Div., Sect. 15-A, February 1965.

[100] D. A. Mortimer and M. Nicholas, J. Mater. Sci., 5, 194 (1970).

THE BEHAVIOR OF FISSION PRODUCTS CAPTURED
IN GRAPHITE BY NUCLEAR RECOIL

Seishi Yajima

Research Institute for Iron, Steel, and Other Metals
Tohoku University
Sendai, Japan

I. INTRODUCTION

In the fuel of a high-temperature gas-cooled nuclear reactor uranium
carbide particles are dispersed in a graphite matrix, which is clad with an
impervious graphite sheath. To develop such a fuel, one of the most important research problems is to establish the technology of minimizing the
release of fission products from the fuel element into the primary gas-cooling
system.

During initial investigations the main objective of many researchers was to obtain the diffusion coefficients of the individual fission products in graphite and to clarify the migration mechanism of the fission products in the graphite matrix. However, the release rates of xenon, krypton, iodine, etc., were high, necessitating the development of pyrolytic-carbon-coated fuel particles. The behavior of the fission products in coated-particle fuels will not be referred to in this chapter, because there appear to be few publications concerned with the relation between fission-fragment damage and fission-product diffusion in pyrolytic carbon materials. Most research in this field is still focused on the development of an adequate technique for making the coated-particle fuel element.

What behavior do the fission products captured in graphite exhibit at high temperatures? What kind of radiation damage do the fission fragments inflict on the graphite crystal? How do the radiation damage and the intrinsic crystal defects in the graphite crystal affect the migration of the fission products in the graphite at high temperatures? Ten years ago these questions had not been clarified.

The fission products are composed of the elements listed in Table 1. Moreover almost all of the nuclides change to other nuclides by successive β decay, so that it becomes very difficult to trace the individual fission products in the graphite. Perplexing phenomena involving fission products are also encountered with the uranium dioxide fuel used in many power reactors today. However, even in this case research on the behavior of fission products at high temperatures was initiated only about 12 years ago.

The crystal structure of uranium dioxide is of the calcium fluoride type, but graphite has a typical layer structure with a distinct structural anisotropy for the a- and c-axis directions. Therefore it is easily conceivable that the migration of the fission products in the graphite crystal along the c-axis should be considerably different from that along the a-axis. Because of this particular characteristic of graphite, it might be interesting to study the behavior of the fission products captured in the graphite crystal by fission recoil. This chapter is a review of a number of investigations related to fission products in graphite.

TABLE 1

Abundances of Fission Products in an Irradiated Uranium Block[a]

Element	Quantity	
	Weight (g)	Volume (ml, NTP)
Kr	44	12, 600
Rb	22	14
Sr	102	39
Y	55	10
Zr	351	54
Mo	298	29
Tc	76	--
Ru	143	12
Rh	26	2
Pd	13	1
Te	39	6
I	17	4
Xe	350	63, 600
Cs	320	170
Ba	109	31
La	110	18
Ce	239	35
Pr	103	16
Nd	345	50
Pm	8	--
Sm	33	4
Eu	2	

[a]Natural uranium, 1 ton, 3000 MWD/ton, 2-MW rate.

II.   FISSION-PRODUCT CAPTURE IN GRAPHITE

In order to capture fission products in graphite, the usual method is to mix graphite powder with sufficient uranium oxide powder and to subject it to neutron irradiation. The fission fragments generated by nuclear fission recoil from the uranium oxide into the graphite crystal and are trapped there. The mean value of the fission-fragment range in the graphite crystal is about 10 μm, although the range depends on the mass of the fission fragments. This compares with a mean value of about 5 μm in uranium dioxide. There-fore, if uranium dioxide powder with a particle size of less than 5 μm, is mixed with a sufficient amount of graphite powder and then irradiated, almost all the fission products generated in the uranium dioxide will be captured in the graphite powder. The relation between the fraction of fission products captured and the weight ratio of uranium dioxide to graphite has been studied theoretically, and the results have been experimentally verified [1].

To eliminate the uranium dioxide from the postirradiated sample, the powder mixture is immersed in a nitric acid solution, since the fission pro-ducts in graphite are scarcely leached out by such a treatment. Nakai et al. [2] investigated the amounts of fission products leached from the graphite by nitric acid solutions of various concentrations and found that the amounts were less than 2% for all elements studied. Findlay and Laing [3] investiga-ted the leaching of the fission products captured in massive graphite samples by 1, 2, and 4 N nitric acid solutions. In their experiment, cylindrical graphite blocks, 1.3 cm in diameter and 1.3 cm in length, were impregnated with molten uranyl nitrate, heated at 400°C to convert the uranyl nitrate to uranium trioxide, and then irradiated. Findlay and Laing concluded that 90 to 95% of the fission products were retained in the graphite sample after the nitric acid treatment. The graphite sample separated from the uranium compound was dried and could be used for experiments on the diffusion of the fission products.

III. BEHAVIOR OF GASEOUS FISSION PRODUCTS IN GRAPHITE

The amounts of krypton and xenon in uranium fission products are large, as shown in Table 1. Furthermore these elements are gaseous even at reasonably low temperatures and do not combine with other materials except in very special circumstances. Because of their inertness, their behavior in graphite can be directly related to the structure of the graphite crystal. On the other hand, the fission products most easily released from the high-temperature gas-cooled fuel into the primary cooling system would probably be these rare gases. All this led to a most careful study of the behavior of the rare gases in graphite as compared with other fission products.

A. Initial Research

At an early stage an extensive study of fission-product diffusion in graphite at high temperatures was carried out by Cubicciotti [4]. In his experiments massive graphite samples impregnated with molten uranyl nitrate were heated at a sufficiently high temperature to convert the uranyl nitrate to uranium carbide. The heat-treated samples were irradiated, and most of the fission products generated by nuclear fission were captured in the graphite. The postirradiated samples were heated isothermally at various temperatures between 900 and 1490°C in a flow of argon. The amount of xenon-133 that diffused from the samples was continuously measured by an ion chamber attached to the experimental system. The remaining xenon-133 was determined by combustion of the samples in an oxygen stream. Some examples of the relation between the square of the evolved fraction f versus the heating time t obtained from this procedure are shown in Fig. 1. In these release curves an initial rapid release can be distinguished from a steady slower release. For the latter region Fick's law of diffusion can be applied since f^2 is apparently linear with respect to t. Cubicciotti [4] further investigated the effect of the size of the graphite specimens on the xenon diffusion.

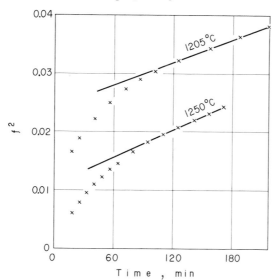

FIG. 1. Xenon-133 release from graphite, f^2 versus time. After Cubicciotti [4].

The apparent diffusion constant D' was found to be independent of the sample size, so that the temperature dependence of the apparent diffusion constants of all specimens used could be fitted to the equation

$$D' = D'_0 e^{-E/RT}.$$ (1)

The values of D'_0 and E are 3.1 sec^{-1} and 49 kcal/mole, respectively.

In discussing his results Cubicciotti intuitively assumed that the migration of xenon would occur only between the basal planes of the graphite crystal, and not across the planes themselves. However, for this process the measured activation energy of 49 kcal/mole appeared to be too high. An explanation of this high value was given in terms of steric hindrance by the graphite layers, because their 3.4-Å separation is smaller than the atomic diameter of xenon, 4.4 Å. This assumption later turned out to be incorrect, because Iwata et al. [5] showed theoretically that the xenon atom should easily migrate between the layers even at liquid-helium temperature.

It is very important to clarify the mechanism of the initial rapid release found in the isothermal heating experiments. Workers at the Argonne National Laboratory [6-9] had also found similar results, and their interpretation of the release curves involved two diffusion mechanism: a rapid "pore" diffusion and a slow "crystallite" diffusion. If there really were a rapid diffusion process that eventually gave way to a flow diffusion rate at a given temperature, then an increase in temperature should not cause a renewed rapid reaction if the amount of rapidly diffusing xenon had already been released at the lower temperature. From Cubicciotti's experiment, however, it was found that with each increase in temperature there was a rapid release, followed by slow diffusion [4]. Cubicciotti could not elucidate the mechanism of the initial rapid release and merely mentioned that the rapid diffusion was governed by some phenomenon that depended on the rate of the temperature rise.

Nakai et al. [10] investigated the release of xenon-133 recoiled into various powder specimens, such as natural, artificial, and amorphous graphites. The initial rapid evolution and the prolonged slower release were recognized in their experiment to be the same as those observed by Cubicciotti [4]. However, the fraction of the initial rapid release during the same heat treatment depended on the type of specimen. The fraction evolved from the natural graphite was overwhelmingly larger than that from artificial and amorphous specimens, as shown in Fig. 2. The mean particle sizes of the natural-graphite specimens used were about 50 and 1 μm. On the other hand, the mean particle sizes of the artificial and amorphous graphites were about 30 and less than 1 μm, respectively. If the initial rapid release had obeyed Fick's law, the amorphous-carbon specimen, which had the smallest particle size, should have shown the largest xenon release. As this was not the case, the initial release cannot be governed by simple diffusion such as described by Fick's law.

Yajima et al. [11] found the same phenomenon with the release of iodine-131 and tellurium-132 from various graphite specimens. Their experimental results are shown in Fig. 3. The natural-graphite specimens were Ceylon graphite and SP-1, the crystals of which were highly graphitic. From Figs. 2 and 3 it is obvious that the fraction of the initial rapid release has a tendency to increase with increasing graphitization.

Findlay and Laing [3] studied the release of the xenon captured by nuclear recoil into various graphite specimens. They too observed the initial rapid and subsequent slower release in the isothermal release curves. In the

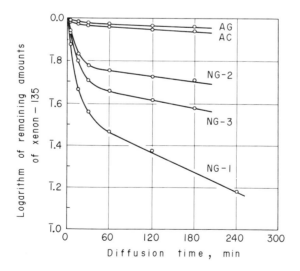

FIG. 2. Xenon-135 release from various graphites. Key: AG, artificial graphite; AC, amorphous carbon; NG, natural graphite.

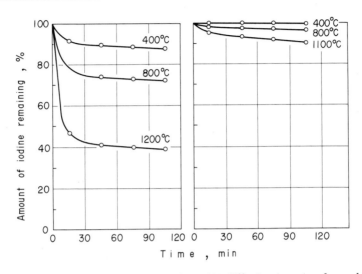

FIG. 3.   Isothermal curves of iodine-133 diffusion in natural graphite (left) and amorphous carbon (right).   After Yajima et al. [11].

region of the slower release rate the fraction evolved was apparently linear with t (t = diffusion time), so that the diffusion constants could be evaluated by using Fick's diffusion equation at various temperatures.   Their results indicated that the activation energy in the region from 500 to 1000°C was different from that above 1000°C, namely, 17 and 53 kcal/mole, respectively.   Findlay and Laing also examined the effect on the diffusion constants of crystal imperfections in the graphite specimens.   However, the activation energies thus obtained from the various specimens were the same only in the range of the slower diffusion.   The initial rapid release was attributed to the easy release of xenon atoms captured in pores and at dislocations, since the xenon generated by fission could easily migrate in the graphite crystal even at low temperatures under irradiation.   It might be concluded that Findlay and Laing lost the chance to clarify the relationship between xenon diffusion and the imperfection of the graphite crystal since the dependence of the fraction of the initial rapid release on the type of graphite was not sufficiently evaluted.

Walker et al [12] also investigated the mechanism of fission-gas diffusion in graphite.   From previous investigations they recognized some dependence of xenon diffusion in graphite on crystal perfection. They therefore used a highly graphitic, pure, natural-graphite powder, SP-1, since the many intrinsic imperfections in a carelessly prepared specimen would have complicated the analysis of xenon diffusion. They found that the release curves were all characterized by a very rapid release in the first part of the heating period, followed by a slower release, and finally by a very slow release that could best be described as a pseudo-plateau. These experimental results were similar to those already described. A second observation concerned the unusual nature of the composite release curve, which was obtained by heating a sample of the graphite until the plateau was reached, withdrawing the sample from the furnace, raising the furnace temperature, inserting the sample again, and heating until a new plateau was reached, etc. Step-release curves of this type showed that the total cumulative release depended only on the maximum temperature and was insensitive to the previous heating history. If on reaching a given plateau the sample was removed from the furnace and then reinserted at the same temperature, there was no rapid release and the release curve followed the original plateau.

Walker et al. [12] also discussed the initial rapid release in connection with the various imperfections in the graphite crystal. Xenon has a low solubility in the graphite lattice, and, after being recoiled into the lattice, it diffuses preferentially to fine pores and dislocations in the crystallite boundaries. Diffusion from the boundaries and escape from the graphite are then characterized by a spectrum of activation energies, which reflect the different crystal imperfections found in the boundaries. At a given temperature only a certain fraction of the xenon atoms in the boundaries has sufficient energy to escape. This explains the rapid release observed at the beginning of a release curve and the subsequent pseudo-plateau.

Morrison et al. [13] studied the release, between 700 and 1500°C, of xenon captured in laminar and columnar pyrolytic carbon coatings on uranium dioxide particles. For these specimens an initial rapid release and a subsequent slower release of xenon could be observed, just as in the usual graphite specimens. The apparent diffusion coefficients for the laminar coatings were 10^2 to 10^3 times higher than those for the columnar coatings with activation energies of 33 and 17 kcal/mole for the laminar and columnar coatings, respectively. Xenon diffusion in pyrolytic-graphite specimens with a quite perfect crystal structure was also investigated. The

release in the a-direction occurred with a higher apparent diffusion coefficient and lower activation energy for temperatures up to 1200°C as compared
with the release in the c-direction. The activation energy for release in
the a-direction was 15 kcal/mole, whereas the value for the c-direction was
50 kcal/mole.

From a number of studies it can be concluded that xenon release from
graphite consists of an initial rapid release and a subsequent slower release.
The fraction evolved during the initial rapid release depends on the crystal
imperfections of the various graphite specimens. It can therefore be assumed that the fission xenon atoms in graphite are bound to the intrinsic and
extrinsic defects of the crystal. The extrinsic imperfections are those produced by fission fragments and fast neutrons during the irradiation. They
can be annealed at lower temperatures than the intrinsic imperfections, which
do not disappear even at the graphitization temperature of about 3000°C.
There is a fair chance of xenon atoms being released from the intrinsic imperfections during heat treatment because of increased lattice vibrations.

B. The Relation between Xenon Diffusion and Radiation Damage

Yajima et al. [14, 15] were the first to observe the influence of crystal
imperfections on xenon release. As already mentioned, Iwata et al. [5]
proved that xenon atoms between graphite layers could easily migrate along
the a-direction in spite of the larger diameter of the xenon atom compared
with the interlayer distance. Yajima et al. concluded therefore that xenon
atoms, captured in the imperfect regions of the graphite crystal, could be
rapidly released if the imperfections that restricted them disappeared during
heat treatment. Some of the liberated xenon atoms would be recaptured in
the remaining imperfections of the crystal, and some of these atoms would
again have the opportunity of being liberated with the aid of thermal vibrations. Such a concept is considerably different from diffusion due to a
concentration gradient of the diffusant. The easy migration of xenon atoms
between graphite layers means that graphite is an extraordinarily suitable
specimen for research on crystal imperfections that restrict the migration
of radioactive xenon atoms.

In this section the relation between xenon diffusion and the intrinsic
imperfection of the graphite crystal will not be explained in detail, because

intrinsic imperfections do not disappear even at a graphitization temperature of about 3000°C. They will therefore still exist during the heat treatments connected with diffusion experiments, which are generally carried
out below 2000°C. As the imperfections are retained during the diffusion
experiment, the xenon atoms bound to them will hardly be liberated from
them. In fact, the fraction evolved from an artificial-graphite specimen is
far smaller than that from a highly graphitic natural-graphite specimen.

Magnuson et al. [16] studied the annealing mechanism of deuteron-
irradiated noble metals. It is assumed that among the various imperfections generated by deuteron irradiation, there is an imperfection that dis-
appears with an activation energy E. If a physical property P of the irradiated specimen (e.g., electrical resistivity as measured in the cited experiment) is affected by this imperfection, its change with annealing temperature
T per unit time can be described by the following equation:

$$\frac{dP}{dt} = -Ke^{-E/kT} \tag{2}$$

where $K = fqA/a = PA/a$, q is the concentration of defects with activation
energy E, f is the contribution factor that describes the amount of property
change associated with an unit concentration of defects, A is an appropriate
vibration frequency, and a is the average number of jumps required for the
moving defects to reach annihilation centers. From Eq. (2) we obtain

$$\frac{dP}{P} = -\frac{A}{a} \exp\left(-\frac{E}{kT}\right) dt, \tag{3}$$

whence

$$P(E) = P_0(E) \exp\left[\left(-\frac{A}{a}\right)\exp\left(-\frac{E}{kT}\right)t\right], \tag{4}$$

where $P_0(E)$ is the initial value of the property at the annealing temperature
T and P(E) is the value after heat treatment for a time t at a temperature T.

If a value of 10^{13} is assumed for A, which is usual for the vibration
frequency, the value of $\exp[-(A/a)\exp(-E/kT)t]$ changes rapidly from 1 to
0 with a minute increase of T in the region of $E = kT$. When a specimen with
a distribution of various defects is heated by a stepwise increase of temperature, the defects with activation energies corresponding to the increasing
values of kT will disappear successively. Therefore the properties that are
affected by those defects will change rapidly. Vand [17] and Primak [18] used

this concept in analyzing the annealing process of imperfections in various crystals. Iwamoto and Oishi [19, 20] also tried to analyze the release of xenon captured by imperfections in a graphite crystal in this way.

For the case of a graphite specimen that contained trapped xenon atoms various peaks of xenon release could be observed when the specimen was heated with a constant increment of temperature. Figure 4, obtained by Yajima et al. [14, 15], shows the release curves of xenon-133 for natural- and artificial-graphite specimens during differential heating at the rate of 5°C/min. With natural graphite many sharp peaks of xenon release can be observed up to 1400°C, but with artificial graphite no peak can be found up to that temperature, at which the release rate of xenon increases rapidly with increasing temperature. This indicates the existence of stronger imperfections in the artificial-graphite specimens compared with the natural-graphite ones, and therefore xenon atoms are bound more effectively by the former. The remarkable difference between the initial rapidly evolved fraction in natural and artificial graphite during isothermal heating can be clearly understood from this experimental result.

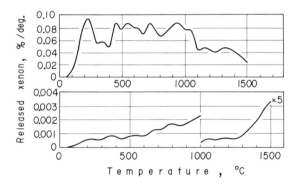

FIG. 4. Xenon-133 release from natural graphite (top) and artificial graphite (bottom) in differential heating, (5°C/min). After Yajima et al. [14, 15].

In general the escape rate V of xenon captured at imperfections in graphite with an annealing activation energy E is given by

$$V = -\frac{dn}{dt} = -\frac{n(x,\ y,\ t)}{a(x,\ y)}\,dx\ dy\ \nu\ \exp\left(-\frac{E_D + E}{kT}\right),\qquad (5)$$

where n(x, y, t) is number of fission xenon atoms captured in a small region between x and x + dx, and y and y + dy; a(x, y) is number of jumps of xenon atoms required to escape from the graphite; E_D is activation energy of xenon-atom diffusion in interlaminar space (this energy is assumed to be very small); E is the activation energy of the disappearance of an imperfection in which xenon atoms are captured; T is the temperature in degrees Kelvin; ν is the vibration frequency of xenon atoms in graphite; and k is Boltzmann's constant.

For a determination of the activation energy, an isothermal excape rate of fission xenon atoms from a sample is first determined at a temperature T_1 and then, after a time t, the temperature is instantaneously increased to T_2 and again an isothermal excape rate is determined. When extrapolated to the time t, the same concentration of xenon atoms can be assumed for both temperatures, and the following equation can be applied

$$\ln \frac{V_2}{V_1} = \frac{E_D + E}{k} \cdot \frac{T_2 - T_1}{T_1 T_2} \tag{6}$$

When the fission xenon atoms are trapped in imperfections, which have a distribution of energy states, Eq. (6) must be carefully applied to the calculation of the activation energies. However, according to Magnuson et al. [16], if the difference between temperatures T_1 and T_2 is made as small as possible, the most important activation energy for the diffusion of xenon atoms within the range of T_1 and T_2 can be calculated from Eq. (6).

Yajima et al. [14] determined the activation energy for xenon release from natural and artificial graphite by the procedure described here. The experimental results are shown in Fig. 5. The activation energy obtained

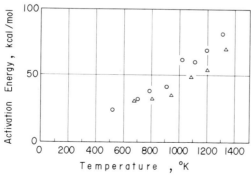

FIG. 5. Apparent activation energy versus temperature. Key: ○ , artificial graphite; Δ, natural graphite. After Yajima et al. [14].

for the temperature range from room temperature to 1450°C cannot be
represented by a single value, but by a continuous distribution ranging from
several kilocalories per mole to about 70 kcal/mole. The values obtained for
both types of specimen (i.e., the natural- and artificial-graphite samples)
are fairly consistent, as seen from Fig. 5. This shows that the kinds of
imperfection that exist in the samples and disappear in the above temperature range are similar. However, as already explained, the release rate
in artificial graphite is much lower, indicating a larger number of imperfections that do not anneal below a temperature of 1450°C. Most of the xenon
atoms in the artificial-graphite specimens were therefore trapped at such
imperfections.

A number of studies on the damage of graphite by fast-neutron irradiation have been reported, but only a few are related to fission-fragment
damage. The primary process of radiation damage due to both neutrons
and fission fragments, however, is the generation of Frenkel pairs. As the
concentration of Frenkel pairs increases, they begin to overlap and are
finally converted to clusters, dislocations, etc. In graphite the number of
Frenkel pairs generated by a fission fragment is several hundred times
larger than that generated by a fast neutron.

Yajima et al. [15] investigated the effect of fission-fragment damage on
xenon diffusion in natural- and artificial-graphite specimens containing
fission fragments of various concentrations. These specimens were heated
from room temperature to about 1500°C at a rate of 5°C/min, and the
xenon-release curves were obtained. Table 2 shows the fast-neutron dose
and the concentration of fission products in the specimens. The release
curves for natural- and artificial-graphite samples are shown in Figs. 6
and 7, respectively. The curves for natural- and artificial-graphite
samples both as irradiated at 500°C and postirradiation annealed are shown
in Fig. 8. Figure 9 represents the curve for the neutron-irradiated natural
graphite. The following remarks may be made on the basis of these figures:

1. Sample 15 was irradiated with the same dose as sample 13, but the
fission-product concentration of sample 15 was greater by a factor of 24
because of the different relative concentrations of uranium oxide and graphite.
The corresponding release curves h and i in Fig. 6 are considerably different from each other. On the other hand, the graphite portion of sample 21
was previously irradiated with a neutron dose 58 times larger than that of
sample 4, but the fission-product concentrations in both these samples were

TABLE 2

Thermal-Neutron Doses and Fission-Product Concentrations in Irradiated Graphite[a]

Sample No.	Graphite type[f]	UO_2 concentration[c] (mg)	Thermal-neutron-dose[d] (n/cm²)	Ratio[b]		T^e (°C)	Fission product concentration
				Initial irradiation	Reirradiation		
1	NG-1	30	1.5×10^{14}	1/60		40	1.8×10^{11}
2	NG-1	30	9.0×10^{14}	1/10		40	1.1×10^{12}
3	AG	30	9.0×10^{14}	1/10		40	1.1×10^{12}
4	NG-1	30	9.0×10^{15}	1		40	1.4×10^{13}
5	AG	30	9.0×10^{15}	1		40	1.4×10^{13}
6	NG-1	30	5.4×10^{16}	6		125	6.4×10^{13}
7	AG	30	5.4×10^{16}	6		125	6.4×10^{13}
8	NG-1	30	2.2×10^{17}	24		125	2.6×10^{14}
9	NG-1	30	8.2×10^{17}	91		125	9.7×10^{14}
10	AG	30	8.2×10^{17}	91		125	9.7×10^{14}
11	NG-1	30	2.6×10^{18}	288		125	3.1×10^{15}
12	AG	30	2.6×10^{18}	288		125	3.1×10^{15}
13	NG-1	30	9.9×10^{18}	1092	∞	125	1.2×10^{16}
14	AG	30	9.9×10^{18}	1092	∞	125	1.2×10^{16}
15	NG-1	1000	9.9×10^{18}	1092	∞	125	2.9×10^{17}
16	AG	1000	9.9×10^{18}	1092	∞	125	2.9×10^{17}

TABLE 2 (continued)

Sample No.	Graphite type[f]	UO$_2$ concentration[c] (mg)	Thermal-neutron dose[d] (n/cm^2)	ratio[b] Initial irradiation	ratio[b] Reirradiation	Te (°C)	Fission product concentration
17	NG-1	1000	9.9 x 10^{18}	1092	8	125	2.9 x 10^{17}
18	NG-1	1000	9.9 x 10^{18}	1092	8	125	2.9 x 10^{17}
19	NG-1	1000	2.4 x 10^{18}	482	4	500	6.0 x 10^{16}
20	AG	1000	2.4 x 10^{18}	482	4	500	6.0 x 10^{16}
21	NG-1	30	5.0 x 10^{17}	58	1	40	1.1 x 10^{13}

[a]Data from Ref. [15].

[b]The flux time was compared with that of sample No. 4, taken as standard.

[c]Milligrams of UO$_2$ mixed with 1 g of graphite.

[d]In every case the doses of fast neutrons were one-tenth of those of thermal neutrons. The thermal-neutron dose was measured with a cobalt monitor, and the fast-neutron dose with a nickel monitor.

[e]Irradiation temperature.

[f]NG-1 and AG represent natural- and artificial-graphite powders, respectively.

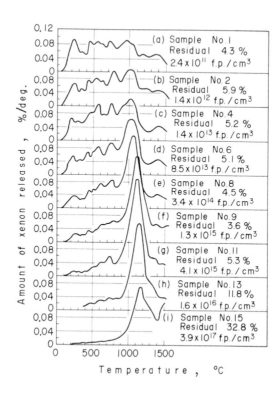

FIG. 6. Xenon-133 release from natural graphite in differential heating, (5 °C/min). After Yajima et al. [15].

almost identical (Fig. 9). From this it can be concluded that the effect of neutron irradiation on the release-curve spectra is very much smaller than that of the fission-fragment irradiation, at least for the dose range in question.

2. The release curves for natural graphite represent a continuous spectrum, possessing several peaks.

3. The positions of the release peaks do not change appreciably with the fission-product concentration. However, returning to Fig. 6, it can be seen that the position of the peak near 1000°C shifts toward higher temperatures with increasing fission-product concentration; that is, the position of

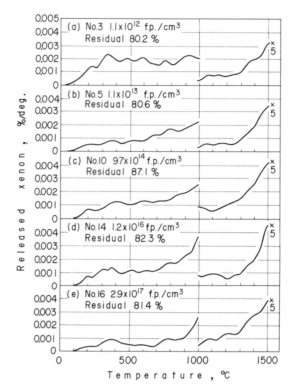

FIG. 7. Xenon-133 release from artificial graphite in differential heating, (5°C/min). After Yajima et al. [15].

the peak is at 1010°C in curve b and finally reaches 1170°C in curve i. This temperature shift might be explained as follows: for a given heating rate the probability of a xenon atom's being captured before it can diffuse out of the graphite increases with the number of imperfections. The concentration of imperfections influences the position of the peak and depends on the fission-product concentration. Another possible explanation is that the trapping power of the imperfections corresponding to this peak increases gradually with fission-product concentration. This temperature shift is not observed with other lower temperature peaks.

4. To determine the influence of the fission-product concentration on the annealing temperatures, three temperature ranges, A (0-850°C), B (850-1350°C), and C (above 1350°C), were chosen arbitrarily. The fraction of the total xenon released in each temperature range is plotted against the

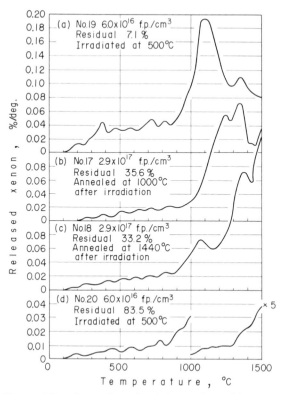

FIG. 8.   Xenon-133 release from heat-treated graphite samples in differential heating (5°C/min).   After Yajima et al. [15].

fission-product (f. p.) concentration in Fig. 10, which shows that the fraction
of xenon released in range A starts to decrease at about 10^{13} f. p. /cm^3,
whereas in range B it continues to increase with the fission-product concentration to reach a maximum at about 10^{15} to 10^{16} f. p. /cm^3.   The released
fraction in range C is constant up to 10^{15} f. p. /cm^3; from there on it increases rapidly.   This appears to indicate that, with an increasing fission-
product concentration, the imperfections in the graphite crystal corresponding
to range A are replaced by imperfections corresponding to range B and finally by those of range C.

5.   As in the case of natural graphite, the release curves of artificial
graphite in differential heating (Fig. 7) were replotted to relate the portion
of the released xenon to the fission-product concentration.   This is shown in

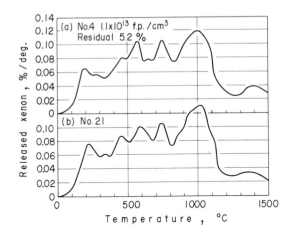

FIG. 9. Xenon-133 release from neutron-irradiated graphite samples in differential heating (5 °C/min). After Yajima et al. [15].

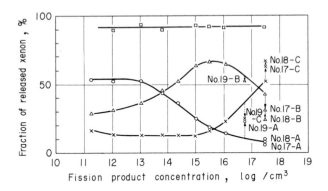

FIG. 10. Change of xenon fraction released from three temperature ranges A, B, and C with fission-product concentration in unit volume of graphite. Key: O, natural graphite, range A (0-850°C); △, natural graphite, range B (850-1350°C); X, natural graphite, range C (> 1350°C); □, artificial graphite, range C (> 1350°C). After Yajima et al. [15].

the uppermost curve of Fig. 10, which represents the fraction of xenon released in the temperature range C only and indicates that the released fraction in C stays at a constant value of over 90% and is thus predominant over those of ranges A and B, which are so small as to be possibly due to experimental error.

6. The curves of the heat-treated samples 17, 18, 19, and 20 (Fig. 8) were also analyzed in the same way and are included in Fig. 10. It is seen that the fraction of xenon released in the temperature range C is greater for samples 17 and 18 than it is for the corresponding nontreated sample, No. 15. From this it can be concluded that postirradiation heat treatment at 1000 or 1440°C will remove the radiation damage corresponding to the ranges A and B, but not to that of range C. It can also be seen that the fission-fragment damage due to reirradiation is very small compared with the residual damage in range C due to the initial irradiation.

In Fig. 10 the curve for sample 19 lies above that of the non-heat-treated samples in range A. This means that performing the irradiation at 500°C impedes the formation of damage that, on annealing, is released in range C.

It can be concluded that postirradiation heat treatment is not effective in removing damage corresponding to higher temperatures. On the other hand heat treatment during irradiation will considerably reduce this damage. From this the following interpretation can be made: The primary radiation damage by a fission fragment is such a slight imperfection that it can be annealed easily by a heat treatment above 500°C. If, however, a graphite specimen is irradiated at low temperature (below 500°C), the generated defects will overlap each other and successively convert to stronger imperfections.

Walker et al. [21] also investigated the relation between xenon release and radiation damage in graphite specimens. Very pure, highly graphitic natural graphite, SP-1, was selected for their experiment to avoid the effect of intrinsic imperfections. They used isothermal heating to obtain the total fractional release of xenon-133 in 30 min (the fraction of rapid release) at three different temperatures, 600, 1000, and 1300°C. The release of xenon-133 is plotted versus the number of xenon-133 atoms per gram of graphite in Fig. 11. The release at 600°C starts to decrease gradually when the xenon-133 concentration exceeds 3×10^{10} atoms per gram. The release at 1000°C decreases in the same concentration range. The release at 1300°C does not show any decrease over the whole range of xenon concentrations investigated. From this experimental result it is clear that the amount of the

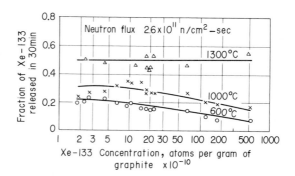

FIG. 11. Fractional release of xenon-133 as a function of xenon-133 concentration. After Walker et al [21].

radiation damage, which anneals during heat treatment between 600 and 1000°C, decreases with increasing concentration of fission-fragment damage. This is qualitatively consistent with the experimental result of Yajima et al. [15]. Walker et al. [21] investigated the influence of the variation of the neutron flux on the release of xenon. The shape and level of the curves in Fig. 11 were independent of the neutron flux (fission-recoil-damage rate) for a fivefold increase in thermal neutron flux. However, for a still higher neutron flux, it has not been clarified whether the xenon release will be affected by the radiation-damage rate or not.

Walker et al. [21] also studied the thermal annealing of heavily damaged graphite, which was then subjected to a short impregnation with xenon-133 to reveal the effect of "old" damage on the diffusion of "new" xenon. Such an experiment had already been performed by Yajima et al. [15], who, however, used a differential-heating method instead of isothermal heating. A sample of graphite impregnated with xenon-133 during a 20-h irradiation at a flux of 1.3×10^{12} n/cm^2-sec was stored at room temperature to allow all the xenon-133 to decay. The sample was then divided into several parts, and these were annealed at one of the following temperatures: 600, 1000, 1400, 1700, 2000, and 2800°C. All the samples were then reimpregnated with a small amount of xenon-133 during a 1-h irradiation at a flux of 2.6×10^{11} n/cm^2-sec. The ratio of the xenon concentration in graphite after a 20- and a 1-h exposure was 80:1. The ratio of the fission-fragment damage was also equal to 80:1.

The results of a 1000 °C diffusion experiment on both the annealed samples (broken curves) and the non-annealed samples (solid curves) are shown in Fig. 12. The annealing temperature is shown on the curves. A comparison with the release curve for a sample that was reimpregnated for only 1 h in a flux of 2.6×10^{11} n/cm^2-sec shows the obvious effect of old damage on the diffusion of new xenon. On the other hand, comparison with the release curve for a sample impregnated with xenon-133 for 20 h in a flux of 1.3 $\times 10^{12}$ n/cm^2-sec indicates that damage is annealed out by thermal annealing. In order to anneal out all the damage a very high temperature is required. The release curves for samples annealed at temperatures between 600 and 1400 °C do not show the expected order (i.e., increasing amount of annealing with increasing temperature), but the reproducibility of the diffusion runs in this temperature range was not good enough to evaluate the effect of annealing temperature on release.

With the object of extending these release experiments to the lowest level of fission-fragment damage possible, a more sensitive technique based on xenon-135 was developed by Walker et al. [22]. The half-life of xenon-135 is much shorter than that of xenon-133, so that the primary fission damage induced by a fission fragment can be investigated precisely. Their

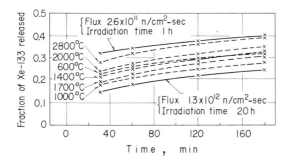

FIG. 12. Effect of thermal annealing of fission recoil damage on xenon-133 released at 1000 °C. After Walker et al. [21].

study could be extended to a damage level at least one order of magnitude lower than that achieved with xenon-133.

The fraction of xenon released in a given time at various temperatures increased with decreasing xenon-135 concentration, that is, with the concentration of fission-fragment damage. It might be possible to obtain some information about the trapped state of xenon-135 in a damaged region from the extrapolated values of the evolved fractions at zero xenon concentration at various annealing temperatures.

Walker et al. [21] investigated the effect of fast-neutron damage on xenon diffusion in graphite. In their experiment graphite specimens irradiated with a dose of 10^{17} nvt fast neutrons were annealed at temperatures from 600 to 2000°C and were impregnated with a small amount of new xenon-133 by fission recoil. It was shown that damage produced by fast-neutron irradiation reduced xenon release. Walker and associates concluded that the effect of fast-neutron damage on the diffusion of xenon was similar to that of fission-fragment damage, but this conclusion was not substantiated by a quantitative analysis.

C. Effect of Intrinsic Imperfections on Xenon Diffusion

Yajima et al. [10, 11] were the first to find distinctly different release rates for xenon, iodine, and tellurium from natural and artificial graphites (Figs. 2 and 3). Their experiments were extended to study the effect of fission-fragment damage on xenon diffusion in graphite single crystals. However, a quantitative relation between the imperfection of the graphite crystal and the xenon-release rate was not established. Findlay and Laing [3] studied xenon diffusion in various graphite specimens: pile grade A powder, natural Ceylon graphite, Magecol 888, and furfuryl alcohol coke. However, they did not discuss the relation between the rapidly evolved fraction and the crystal perfection of the various specimens. The release rate of xenon from furfuryl alcohol coke was remarkably high in comparison with that from the other specimens and was attributed to the highly porous nature of the material. Furfuryl alcohol coke is similar to glassy carbon, its crystal structure being extremely nongraphitic. It is noteworthy that the release rate of xenon from such crystals is higher than that from natural graphite in spite of the former's being somewhat amorphous. Yajima et al. [23] recently investigated the release rate of xenon from pyrolytic-graphite specimens prepared at about 1700°C, in which the crystallite orientation was extremely random and found that the release rate was much higher than that found for the normally oriented pyrolytic graphite.

Walker et al. [12] investigated the relation between the imperfection of various heat-treated petroleum cokes and the release of xenon from them. Petroleum coke, calcined at 1250°C, was ground and sieved to obtain powders with various particle sizes. The powders were graphitized at various temperatures up to 3000°C. The crystallographic parameters of the calcined coke and of the heat-treated samples of this calcined material are shown in Table 3. High values of the (002) peak intensity indicate a preferred orientation of the basal plane parallel to the faces of the flakelike particles.

The samples were impregnated with xenon-133 by fission recoil, and the xenon-release curves were then obtained by isothermally heating at 800, 1100, and 1400°C. The highest release temperature studied, with the sample carbonized at 1250°C, was 1100°C, to avoid further thermal changes in structure. Several release curves for the 3000 and 1250°C samples are

TABLE 3

Crystallographic Parameters of Graphitized Carbons[a]

Sample	T^b (°C)	d Spacing (Å)	c_L [c] (Å)	X-ray density (g/cm^3)	(002) Relative intensity
1	1250	3.51	34	2.17	17
2	2000	3.42	226	2.225	57
3	2250	3.373	597	2.256	87
4	2500	3.366	852	2.260	93
5	2750	3.360	962	2.265	96
6	3000	3.357	1015	2.267	105

[a]Data from Ref. 12 .

[b]Heat-treatment temperature.

[c]Crystallite height.

shown in Fig. 13, from which it can be seen that they all have the same general shape. However, there is a radiation-damage effect that is qualitatively the same as for SP-1 graphite, but much smaller in magnitude.

The xenon release increased with decreasing particle size, but this particle-size effect was not large. It was also noted that xenon diffusion in the various heat-treated cokes did not obey Fick's law. The most significant feature of the curves in Fig. 13 is the pronounced insensitivity of the total xenon-133 release on the coke structure, showing that large changes in crystallite size and graphitization can occur without great influence on xenon release. It was concluded from the experimental results that diffusion was not sensitive to the structural factors associated with the crystallites and hence that the nongraphitic boundaries between the crystallites are of major importance in the diffusion kinetics.

Using differential heating, Yajima et al. [23] obtained xenon-release curves from pyrolytic-graphite specimens with various d spacings. For a considerably graphitized specimen the xenon-release peak near 1100°C appeared to be the same as for natural graphite (Fig. 4). This suggests that materials of intermediate perfections between natural and artificial graphites can be obtained by heat-treating as-deposited pyrolytic graphite. It might also be possible to elucidate the trapped state of xenon in an intrinsic imperfection by employing various pyrolytic-graphite specimens, in which the structure changes continuously from grossly imperfect to a nearly single-crystal type.

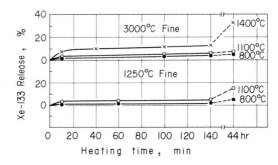

FIG. 13. Xenon-133 release from heat-treated petroleum cokes. After Walker et al. [12].

IV. BEHAVIOR OF NONGASEOUS FISSION PRODUCTS

It is very troublesome to investigate the migration of nongaseous fission products in graphite as compared with that of krypton and xenon. This is because they can combine with graphite at high temperatures, and there is a possibility that their evaporation from the graphite surface is the rate-controlling factor for their release. Furthermore nongaseous fission products have a tendency to deposit on the wall of the experimental apparatus, so that it is necessary to be extremely careful to capture all the fission products diffusing out of the specimen.

On the other hand, the release of iodine, tellurium, and other nongaseous fission products is a more important problem for the development of fuel for high-temperature gas-cooled reactors than that of krypton and xenon. Their contamination of the primary coolant system results in maintenance problems for the whole system, and, in general, nongaseous fission products are important factors from a safety aspect.

Based on the amount of gamma activity released as high-energy decay products with long half-lives, decay chains 87-95, 129, and 131-143 are the most significant. Examination of these decay chains indicates that rubidium, yttrium, strontium, zirconium, and niobium are the most important radioactive products of chains 87-95; antimony, tellurium, and iodine of chain 129; and antimony, tellurium, iodine, xenon, cesium, barium, strontium, lanthanum, and cerium of chains 131-143. Fission chains 147-161 produce rare earths with high neutron-absorption cross sections. The rate of removal of these elements will influence the neutron economy of the reactor.

A. Initial Research

The initial research for nongaseous fission products in graphite was carried out mainly by scientists from the North American Aviation Laboratories [4, 24-26]. In their experiments massive artificial-graphite specimens, such as AGOT, impregnated with molten uranyl nitrate were heated at 1850°C under vacuum to convert the uranyl nitrate to uranium carbide. The treated specimens were irradiated and subsequently subjected to diffusion experiments for fission products in the temperature range 1500 to 2200°C.

In these specimens only about 65% of the fission products was captured in the graphite by fission recoil, so that the release of the fission products originated both from the graphite and the uranium carbide, and consequently the experimental results became rather complicated. Moreover some part of the uranium carbide was converted to uranium oxide during storage of the specimens, and the oxide was reconverted to uranium carbide by the heat treatment of the diffusion experiment. In the event of a structural change in the matrix containing the fission products, the release of the fission products was accelerated. The effect of various pores in the specimens on the release of fission products could also not be ignored. Because of these factors some uncertainties entered the experimental results. It was therefore difficult to clarify the diffusion mechanisms of the individual fission products in graphite from the experimental results, since there was considerable scatter, and the experimental data were insufficient.

Young and Smith [25] showed that more than 99% of barium and strontium were released from graphite during a heat treatment of 5 min at 2100°C, whereas only about 2% of zirconium was evolved during 4 h at 2200°C. They surmised that the appreciable difference between the evolved fractions was due to the different volatilities of the carbides formed in the graphite.

Doyle [26] investigated the release of many kinds of fission product from graphite in the temperature range of 1500 to 1900°C. His results confirmed that the evolved fractions of iodine, tellurium, cesium, strontium, and barium were quite large, that those of ruthenium and rhodium were smaller, whereas molybdenum and ziroconium scarcely diffused from the specimens at all. The release curves of iodine, tellurium, barium, strontium, and cesium obtained in an isothermal-heating experiment showed two regions: an initial rapid release and a lasting slower release similar to that described for xenon. In particular the cesium-release curve could be clearly divided into these two regions, and the initial rapidly evolved fraction was quite large. Doyle considered that the major part of the recoiled cesium could easily migrate between the graphite layers without "steric hindrance" since the diameter of the cesium atom is relatively small (3.3 Å). According to his explanation the slow release process is due to one mode of physical entrapment. However, the behavior of the remaining entrapped cesium still has to be considered as problematic.

Yajima et al. [11] also studied the release of iodine and tellurium. In their experiment highly graphitized natural graphite powder, such as SP-1, less graphitized artificial-graphite powder, and amorphous-carbon powder were employed. The fractions of the initial rapid release of iodine and tellurium from the natural graphite were much larger than those from the

artificial graphite and the amorphous carbon (Fig. 3). The experimental
results suggest that the release of iodine and tellurium in graphite is closely
related to the crystal structures of the specimens.

Cowan and Orth [27] extended the type of diffusant in graphite to the shorter lived nuclides, using a pneumatic tube. For isothermal-heating experiments
in the temperature range 1600 to 2600°C the following elements were investigated: arsenic, bromine, strontium, zirconium, ruthenium, rhodium, palladium, silver, cadmium, tin, antimony, tellurium, iodine, cesium, barium,
cerium, neodymium, and europium. The diffusion coefficients of the individual
elements were obtained from the fractions evolved and the diffusion temperatures. Some typical diffusion losses at 2400°C are shown in Fig. 14, from
which it is clear that the majority of the fission products show an initial rapid
release and a subsequent slower release. The diffusion coefficients and the
corresponding activation energies of the individual fission products were obtained from the slower release region. However, there was no simple correlation between the diffusion rates of the fission products and some simple

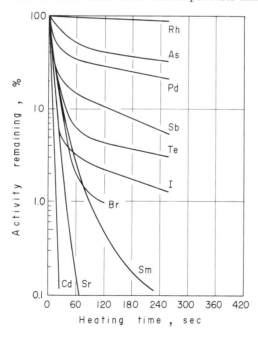

FIG. 14. Some typical diffusion losses in graphite at 2400°C. After
Cowan et al. [27].

property, such as melting or boiling point of the metal or the carbide, or the diameter of the atom or of the probable ion. It was obvious, however, that zirconium and molybdenum, which form stable carbides, were hardly released from the graphite specimen at all. On the other hand, cadmium and tin, which presumably existed in the metallic state and have low melting and boiling points, evolved very rapidly. It was surmised that the remarkably rapid release of silver was due to its low melting point and its small atomic diameter.

Orth [28] studied the release of fission lanthanides from graphite in the temperature range 1600 to 2600°C. The experimental results showed that the release rates increased with decreasing boiling points. The same tendency could be observed with the rates at which the actinide elements were released from graphite. However, it cannot be concluded from these results that the elements considered migrate in an elemental form in the hot graphite.

Saunders [29] studied the diffusion behavior of barium, yttrium, zirconium, and strontium in AGOT graphite in the temperature range 800 to 1800°C. In this case the diffusing elements were the oxides labeled with radioactive isotopes instead of the fission products. An attempt was made to correlate the activation energies of the diffusants with their atomic diameters and melting points, but no clear conclusion could be drawn.

Large and Walton [30] reported that the logarithms of the diffusion constants of various nongaseous fission products in graphite are linearly related to their ionic diameters. This interpretation was based on the random-walk theory for a nongaseous fission product in graphite.

Findlay and Laing [3] investigated the release of solid fission products at 800°C from massive graphite specimens. The apparent diffusion constants obtained for the various elements were as follows (in sec^{-1}): tellurium 5.8 x 10^{-11}, iodine 1.1 x 10^{-11}, cesium 1 x 10^{-13}, cerium and lanthanides 5.7 x 10^{-14}, strontium 1.1 x 10^{-14}, barium 5 x 10^{-16}, niobium and zirconium 3.1 x 10^{-16}. Generally the apparent diffusion constants increase with decreasing boiling point, and the tendency is thus similar to the results previously reported.

Bryant et al. [31] investigated the release of fission products from graphite at temperatures between 1100° and 2500°C. The most interesting result of their experiment concerns the shapes of the release curves of tellurium and barium in isothermal heating, which are much different from each other, as shown in Fig. 15. The rate of barium release remained approximately proportional to the barium concentration at all stages of release. The rate of tellurium release, however, was initially equal to that of barium but then decreased rapidly with decreasing tellurium concentration. The shape of the curves for barium was also typical of strontium, yttrium, cesium, and the lanthanide elements. The rate-limiting process of their diffusion was considered to be diffusion through the graphite matrix. The curve for tellurium was also typical of antimony and iodine. In this case the rate-limiting process was assumed to be an evaporation process, rather than diffusion through the graphite matrix. In this experiment the shapes of the retention-versus-time curves for krypton, rubidium, zirconium, molybdenum, technetium, ruthenium, rhodium, palladium, silver, cadmium, indium, tin, and xenon were not established. It was, however, clarified that silver, cadmium, and tin diffused rapidly even at 1200°C, whereas zirconium, molybdenum, and ruthenium were released only very slowly at 2400°C.

In spite of the great amount work related to the release of the solid fission products in graphite, it is difficult to elucidate the diffusion mechanisms for

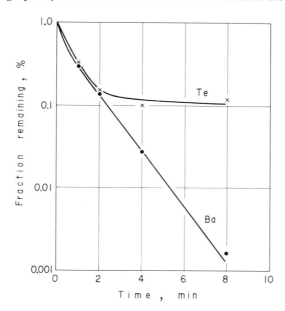

FIG. 15. Losses of barium and tellurium from admixture samples as a function of time at 2400°C. After Bryant et al. [31] .

the individual fission products because of scatter in the measured values and lack of experimental data. However, it can be concluded qualitatively that the release rate of fission products increases with decreasing melting or boiling point and that elements that form stable carbides tend to migrate slowly. It will be impossible to clarify the diffusion mechanisms of the solid fission products in graphite merely by studying the release curves of the individual fission products in isothermal heating. To resolve the situation it is necessary to develop new experimental methods in order to establish the chemical form of the fission products diffusing from the graphite.

B. The Chemical Form of Iodine Diffusing from Graphite

Yajima et al. [32-34] studied the chemical form of the iodine that diffuses out of graphite, to clarify the diffusion mechanism of the fission iodine recoiled into graphite.

In their experiment [32] the following samples were used:

1. Uranium dioxide powder containing fission products.

2. A mixture of natural-graphite powder and uranium dioxide containing fission products.

3. Natural-graphite powder containing fission products (uranium dioxide was removed by a nitric acid treatment).

4. Thorium dioxide containing fission products (fission products were generated by the (γ, f) reaction using a linear electron accelerator).

5. Elemental iodine (I_2) labeled with iodine-131.

The apparatus used is shown schematically in Fig. 16. A quartz tube (0.6 cm in diameter, 100 cm long) was packed with natural-graphite powder, and the samples were placed on top of the powder. The whole apparatus was kept in a purified argon atmosphere. The tube was then heated with external resistance wires so as to produce along its total length a linear temperature gradient of about 14°C/cm ranging from 1000°C to room temperature. After attaining equilibrium the sample was heated for 2 h at 1000°C while a carrier gas of purified argon was made to flow at a rate of 0.5 ml/min. The fission iodine-131 that diffused out of the sample entered the argon stream and was deposited on the graphite powder in the tube. The sample was then cooled in a stream of purified argon, after which the specimen tube was surveyed along its whole length in order to locate the iodine-131 activity. A γ-ray scintillation spectrometer connected to a sodium iodide crystal with a lead slit 1 cm wide was used.

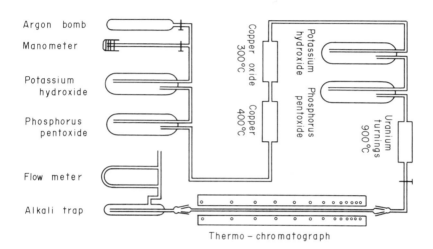

FIG. 16. Thermochromatograph. After Yajima et al. [32].

As shown in Fig. 17, fission iodine from samples a, b, and c deposited at a position corresponding to about 580°C, fission iodine from thorium dioxide (sample d) was found at 520 and 430°C, and elemental iodine (sample e) at about 80°C. The thermochromatograms of the fission iodine-131 were extremely different from that of elemental iodine. The chemical form of the deposited iodine was determined chemically by using various carriers, such as I^-, I°, IO_3^-, and IO_4^-. From this result it was determined that the chemical form of the iodine deposited at 580, 520, and 430°C was the iodide (I^-).

Subsequently Yajima et al. [34] compared the thermochromatogram of the deposited iodine diffusing from graphite with that of uranium tetraiodide labeled with iodine-131. The uranium tetraiodide sample was put into a breakable silica capsule (see Fig. 18), which was filled with an inert gas and sealed because the compound was very unstable in the presence of oxygen and moisture. The carrier gas in the thermochromatograph was circulated by an automatic Toepler pump, as shown in Fig. 19. The apparatus had a bypass gas-purification system, so that it was possible to conduct the experiments with highly

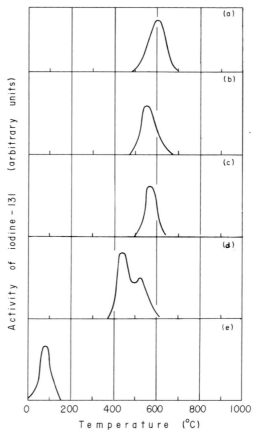

FIG. 17. Thermochromatograms of iodine. After Yajima et al.[32].

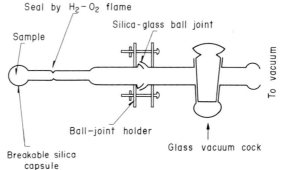

FIG. 18. Assembly of sample holder for thermochromatograph. After
Yajima et al.[34].

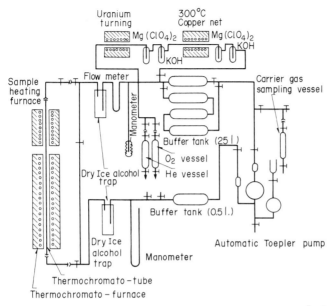

Uranium   300°C
turning   Copper net

Mg(ClO₄)₂   Mg(ClO₄)₂
                                    KOH
KOH

Sample   Flow meter   Carrier gas
heating   sampling vessel
furnace

Manometer

Buffer tank (25 l)

O₂ vessel
Dry Ice alcohol   He vessel
trap

Buffer tank (0.5 l.)

Dry Ice   Automatic Toepler pump
alcohol
trap   Manometer

Thermochromato-tube
Thermochromato-furnace

FIG. 19. Thermochromatograph. After Yajima et al.[34].

purified helium carrier gas. The sample in the breakable silica capsule and
its breaker (an iron bar sealed in a silica capsule) were located in the center
of the sample-heating furnace on top of the packing material of the thermochromatograph tube.

The thermochromatograms of the fission iodine-131, which diffused out
of the graphite sample, and that of the uranium tetraiodide labeled with iodine-
131 are shown in Fig. 20. The deposition temperatures of both samples are
the same. Yajima et al. [34] also investigated the chemical behavior of the
fission iodine and the uranium tetraiodide, which deposited at about 550°C
on the thermochromatograph tubes, under various partial pressures of oxygen
added to the circulated helium stream. The experimental result is shown in
Fig. 21. With increasing oxygen content the position of both deposits shifted
from an initial value of 550°C through intermediate temperatures of 400, 300,
and 200°C. They were finally converted to elemental iodine at about 80°C.
From the similar behavior of both deposits toward oxygen, it seems reasonable
to assume that the chemical form in which fission iodine-131 diffuses out of
graphite is identical with that with which it leaves uranium tetraiodide. However, it was very difficult to detect the uranium in the deposit due to the fis-

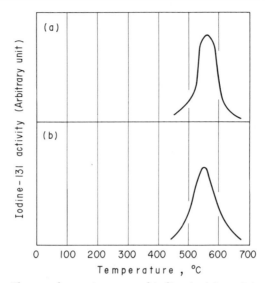

FIG. 20. Thermochromatograms of iodine in (a) a mixture of graphite
and uranium dioxide and (b) uranium tetraiodide. After Yajima et al. [34].

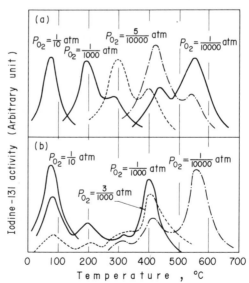

FIG. 21. Thermochromatograms of iodine under various oxygen partial
pressures in (a) a mixture of graphite and uranium dioxide and (b) uranium
tetraiodide. After Yajima et al. [34].

sion iodine-131, since its amount was extremely minute. Recently Handa and
Shiba [35] obtained the molecular weight of the fission iodine-131 deposit that
diffused out of graphite by measuring the diffusion rate of the gaseous molecules of the iodine-131 deposit through minute pores. They showed that it is
identical with the molecular weight of uranium tetraiodide.

It had been observed by Ershler and Lapteva [35] and Rogers and Adam
[37] that, when a foil of uranium metal was irradiated with neutrons under
high vacuum, a large number of uranium atoms evaporated from the foil.
Bierlein and Mastel [38] also recognized the same phenomenon in a uranium
dioxide specimen. Therefore it may be conceivable that, in the experiments
of Yajima et al. [32-34], a considerable number of uranium or thorium atoms
were brought into the graphite crystal by primary fission fragments and that
the fission iodine-131 would be trapped in the imperfection consisting of
unstable atoms of uranium or thorium caused by fission. Uranium or thorium
iodide would then be formed and would diffuse out of the sample.

Yajima et al. [33] obtained curves of the release of iodine-131 from
graphite by using differential heating at a rate of 5°C/min. Argon was used
as the carrier gas, and the flow path, made of a quartz tube, was wound with
a resistance heater to avoid deposition of iodine on the wall. A silver-plated
copper net was used for trapping the iodine. It was designed so as to be
easily exchangeable. An example of the curves obtained in this experiment
is shown in Fig. 22.

It is believed that after diffusion through the sample iodine-131 interacts
physically or chemically with the graphite crystal on the surface before it
leaves. It would therefore be useful to obtain the release curve of iodine-131
that, after diffusion from the graphite crystal, had deposited on the surface
of graphite powder of a similar nature. For this purpose an iodine sample
was prepared by means of the thermochromatographic method already described and put on the surface of the graphite. The release curve of iodine-
131 obtained is also shown in Fig. 22.

In Fig. 22 the release curve of deposited iodine has a peak at 530°C,
while the peak of the release curve of "diffusion iodine" is found at the higher
temperature of 750°C. This leads to the conclusion that the iodide, which had
diffused in graphite and had reached the surface, did not stay there but was
immediately carried off by the argon gas stream; that is, the peak at 750°C
originated in the actual diffusion of iodine-131. It was not established, how-

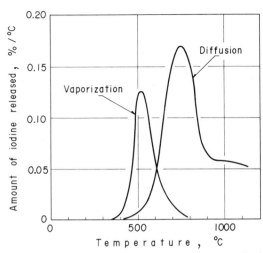

FIG. 22. Iodine-131 release in differential heating (5°C/min). After Yajima et al. [33].

ever, how the release curve in a differential-heating experiment will appear for the remaining iodine in graphite after extending the heat treatment up to about 800°C.

Several isothermal-heating curves in the neighborhood of 750°C, at which the iodine release peak had appeared in the differential-heating curve, were measured [33] . In order to determine the activation energy of the diffusion of iodine-131 in graphite the isothermal release curves obtained were analyzed according to the method of Dienes and Parkins [39]. From the result it was determined that the activation energy of iodine release was about 47 kcal/mole at about 750°C.

C. The Chemical Form of Tellurium Diffusing from Graphite

Shiba et al. [40] used a thermochromatograph to study the chemical form of tellurium diffusing from graphite. In their experiment the alundum thermochromatograph tube was heated by a high-frequency induction coil to produce a linear temperature gradient from 1500°C to room temperature. The tellurium from the graphite specimen was located at a place corresponding to about 1000°C in the thermochromatograph tube. Elemental tellurium labeled with tellurium-132 was examined by the same procedure as that used for the fission tellurium. As shown in Fig. 23, the elemental tellurium was also

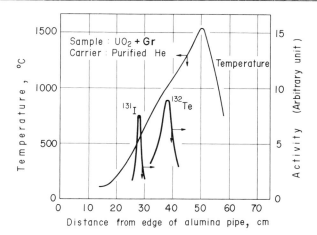

FIG. 23. Thermochromatogram of fission tellurium released from uranium dioxide + graphite in a helium atmosphere. After Yajima et al. [40].

located at about 1000°C. The effect of oxygen on the thermochromatogram of fission tellurium-132 was such that with increasing partial pressure of oxygen in the thermochromatograph, the fission tellurium decreased its deposition temperature from the initial 1000°C until finally it deposited completely at 550°C. The most probable species of the deposit was considered to be an oxide, of which tellurium dioxide is the most stable at high temperatures since tellurium trioxide decomposes largely into TeO_2 and O_2 above 360°C [41]. The thermochromatographic behavior of tellurium dioxide and of tellurium metal was studied on the basis of this assumption. The tellurium dioxide labeled with tellurium-132 deposited at 550°C, as shown in Fig. 24. From these experiments it could be concluded that tellurium-132 diffuses from graphite in the elemental state.

By analogy with the fission iodine, it can be surmised that fission tellurium released from uranium dioxide forms a U-Te compound, though from speculation alone there still remains the other possibility that it forms an oxide. Had the fission tellurium combined with uranium atoms, the deposition temperature should have been above 1200°C, which is reported to be the melting point of uranium ditelluride [42], whereas combination with oxygen should have brought the deposition temperature down to 550°C. Actually the fission tellurium was deposited at 1000°C, corresponding to metallic tellurium.

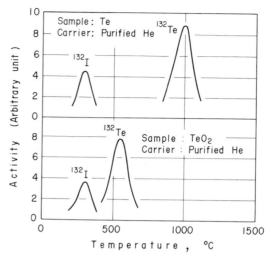

FIG. 24. Thermochromatograms of tellurium and tellurium dioxide tagged with tellurium-132. After Yajima et al. [40].

This behavior can be explained as follows: Since the free energy of formation of uranium ditelluride is much lower than that of uranium tetraiodide, there would be a correspondingly smaller possibility for the fission tellurium to combine with uranium atoms produced in fission spikes by the dissociation of uranium dioxide. The free energy of formation of tellurium dioxide is not definitely known for temperatures near 1000°C, but is believed to be slightly below zero. This would impede the formation of oxide by the fission tellurium.

V. RELEASE OF XENON AND IODINE CAPTURED IN A GRAPHITE CRYSTAL

It is reasonable to consider from the many experimental results that xenon atoms recoiled into graphite are trapped in various imperfections of the crystal, and that they are only liberated and released from the specimen if the imperfections are removed by heat treatment. According to previous experiments [14, 15] many kinds of imperfections in graphite crystals are indirectly indicated by the curve of xenon release in differential heating, and the activation energy of their disappearance does not show a single value but a spectrum. However, it is not known what kinds of imperfections correspond to each activation energy. Therefore the number of xenon atoms trapped in each type of imperfection and the physical condition of the xenon atoms in the imperfections are not clear. Observations of fission-fragment damage and of dislocation configurations in irradiated graphite specimens have been made by a number of workers using electron microscopy, the results have not produced any information concerning the trapped state of xenon in the graphite crystal.

It is interesting to compare the differential-heating curves of the volatile fission products xenon and iodine in graphite with the annealing curves of graphite irradiated with fast neutrons. To depict neutron damage in artificial graphite, differential-annealing curves of several physical properties are shown in Figs. 25 and 26, which are based on work by Woods et al. [43] and by Kinchin [44]. Table 4 lists the activation energies for the annealing of the electrical resistivity, determined by a pulse-annealing technique.

The peaks of the diffusion curve of xenon in isothermal heating and those due to annealing of neutron damage coincide with each other except for the one at 750°C; this coincidence is the more remarkable when one considers the difference in such factors as irradiation conditions, experimental method, and evaluation of the data. This indicates that the xenon-tracer technique can match other methods with respect to high sensitivity and simplicity in studying the process

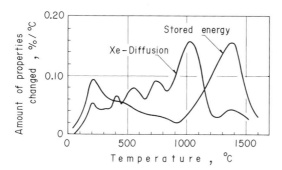

FIG. 25. Xenon release in differential heating (5°C/min) and differential-annealing curve of neutron damage. After Yajima et al. [33].

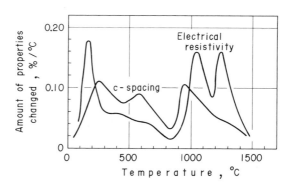

FIG. 26. Differential-annealing curves of neutron damage. After
Yajima et al. [33].

of the annealing of imperfections in graphite crystals. Furthermore the
activation energies of the diffusion of xenon at 200 and 1000°C are nearly
equal to those obtained from the annealing of electrical resistivity. This
means that, so far as these peaks are concerned, the damage due to fast
neutrons is the same as that due to fission fragments and that xenon is released on annealing out these imperfections.

It has been established that fission-fragment damage caused by the recoil of fission fragments into graphite crystals is far greater than the fast-
neutron damage [45]. The absence of a peak at 750°C in the annealing curves
of the neutron damage would indicate either that this damage is characteristic
of fission-fragment damage or that other methods less sensitive than the
xenon-tracer technique do not reveal the presence of the 750°C peak in the
specimens irradiated with fast neutrons. However, the latter suggestion
cannot be easily confirmed. Therefore the imperfection corresponding to
the 750°C peak may have its origin in the fission fragments themselves or in
the fissionable materials, which in the present case are uranium atoms
brought into the graphite crystal by fission fragments. The physical interaction of fission fragments with graphite atoms is not considered to be fundamentally different from that of fast neutrons.

These assumptions lead to the conclusion that iodine-131 was initially
distributed evenly over the various imperfections in the graphite crystal,
just as was the recoiled xenon. However, because of the chemical affinity

TABLE 4

Activation Energies for the Annealing of Irradiation Damage

Property	Exposure	Activation energy (kcal/mole)				Refs.
		150–250°C	500–600°C	650–850°C	1000–1200°C	
Electrical resistivity	8.1×10^{18} n/cm^2	23.8			69.5	43
	1.5×10^{19} n/cm^2	29.2			58.2	43
	5.4×10^{20} n/cm^2	39.6				43
	$\sim 10^{19}$ n/cm^2	27.6			69	
Xenon diffusion	1.4×10^{13} f.p./cm^3	23.8	32.0	48.6	68.4	14,15
	1.4×10^{13} f.p./cm^3				65.5	14,15
Iodine diffusion	2.6×10^{14} f.p./cm^3			47		33

of iodine, iodine-131 that was trapped at imperfections (C_2, C_3, C_4, etc.
clusters) [46] corresponding to temperatures below 750°C had a very low
probability of migrating between layers or of diffusing out of the graphite
crystal and therefore rather tended to combine successively with clusters
annealing at higher temperatures. Thus, as smaller clusters disappear at
lower temperatures, the iodine-131 migrates to larger clusters, which do
not anneal until higher temperatures, and is trapped there. Consequently,
in the case of iodine, no peak is observed at temperatures lower than 750°C.
Finally, iodine-131 combines with impurity atoms, which constitute imperfections corresponding to the peak at 750°C, to form an iodine compound and
diffuses out. This assumption is supported by the fact that at a concentration
of 2.6×10^{14} f.p./cm^3 the amounts of iodine and xenon diffusing at temperatures below 900°C are 42 and 44%, respectively, that is, are nearly equal.
Other evidence is that many halides of metals like UCl_4 and UO_2Cl_2 inter-
calate graphite in the vicinity of 300°C [47]; this quite low temperature facil-
itates the diffusion of the iodide compound in the graphite crystal. The confirmation of the presence of uranium atoms and the determination of the
concentrations of the atoms along the range of fission fragments in graphite
would further consolidate this assumption.

It is necessary to study the behavior of the released iodine above about
800°C by obtaining the differential-heating curve. However, as shown in
Fig. 22, the iodine-131 released from graphite, which is heated at 1500°C,
is deposited at 550°C in the thermochromatograph tube. This suggests that
the chemical form of the iodine diffusing from graphite above 800°C is
uranium tetraiodide, the same as that of the released iodine, which shows
the 750°C peak in the differential-heating curve, as already explained.

VI. CONCLUSION

The difference between natural graphite and the normal artificial
graphites cannot be determined from the usual criteria, such as X-ray diffraction patterns and electrical properties. In spite of their similarity, the
behavior of xenon release from natural graphite is remarkably different
from that released from artificial graphite. This signifies that the use of
radioactive xenon is an extremely sensitive technique in the determination
of graphite perfection, and at the same time the behavior of xenon in graphite
crystals is intimately connected with graphite microstructure. It would be

valuable to prepare specimens with crystal perfections of an intermediate type between natural and artificial graphite and to study their xenon release, because it would help elucidate the state of the trapped xenon in the various imperfections of the graphite crystal.

The study of the behavior of solid fission products is more difficult than the study of xenon. The thermochromatographic technique developed by Yajima et al. [32] is favorable for the study of the volatile fission products like iodine and tellurium, but it is useless for fission products, such as zirconium and niobium, which form stable carbides in the graphite matrix. It is hoped, therefore, that it will be possible to develop an adequate experimental method that will cover the physical and chemical properties of the individual fission products.

ACKNOWLEDGMENTS

The author wishes to express his appreciation to Drs. K. Shiba and M. Handa of the Japan Atomic Energy Research Institute, Tokai, Ibaraki, Japan, who studied the release of the fission products from graphite with the author and whose experimental results have been frequently used in this chapter.

REFERENCES

[1]   E. Nishibori, R. Ueda, S. Yajima, S. Ichiba, Y. Kamemoto, and
      K. Shiba, Powder Metallurgy in the Nuclear Age, Springer-Verlag,
      Berlin, 1961, pp. 198-230.

[2]   T. Nakai, S. Yajima, K. Shiba, J. Osugi, and D. Shinoda, Bull. Chem.
      Soc. Japan, 33, 494 (1960).

[3]   J. R. Findlay and T. F. Laing, J. Nucl. Mater., 7, 182 (1962).

[4]   D. Cubicciotti, NAA-SR-194 (1952).

[5]   T. Iwata, E. Fujita, and H. Suzuki, J. Phys. Soc. Japan, 16, 197
      (1961).

[6]   J. E. Wilson and O. C. Simpson, Summary Report for July, August,
      and September 1946, ANL-4000, Sect. 1.5, June 2, 1947, pp. 56-57.

[7]   D. Schultz, Summary Report for January, February, and March 1947,
      ANL-4006, Sect. 1.6, October 3, 1947, pp. 62-63.

[8]   D. Schultz and O. C. Simpson, Summary Report for April, May, and
      June 1947, ANL-4090, Sect. 1.6, January 6, 1948, pp. 28-57.

[9]   D. S. Gaarder and O. C. Simpson, Summary Report for July through December 1947, ANL-4185, Sect. 1.5, October 1, 1948, pp. 25-38.

[10] T. Nakai, S. Yajima, K. Shiba, J. Osugi, and D. Shinoda, Bull. Chem. Soc. Japan, 33, 497 (1960).

[11] S. Yajima, S. Ichiba, Y. Kamemoto, and K. Shiba, Bull. Chem. Soc. Japan, 34, 493 (1961).

[12] P. L. Walker, Jr., H. B. Palmer, and W. S. Diethorn, TID-19896 (1963).

[13] D. L. Morrison, T. S. Elleman, R. H. Barnes, and D. N. Sunderman, BMI-1634 (1963).

[14] S. Yajima, S. Ichiba, Y. Kamemoto, K. Shiba, and M. Kori, Bull. Chem. Soc. Japan, 34, 697 (1961).

[15] S. Yajima, S. Ichiba, K. Iwamoto, and K. Shiba, Bull. Chem. Soc. Japan, 35, 1263 (1962).

[16] G. D. Magnuson, W. Palmer, and J. S. Koehler, Phys. Rev., 109, 1990 (1958).

[17] V. Vand, Proc. Phys. Soc. (London), A55, 222 (1954).

[18] W. Primak, Phys. Rev., 100, 1677 (1955).

[19] K. Iwamoto and J. Oishi, J. Nucl. Sci. Tech., 4, 223 (1967).

[20] K. Iwamoto and J. Oishi, J. Nucl. Sci. Tech., 4, 431 (1967).

[21] P. L. Walker, Jr., H. B. Palmer, and W. S. Diethorn, TID-21415 (1964).

[22] P. L. Walker, Jr., H. B. Palmer, W. S. Diethorn, and D. E. Kline, NYO-1710-66 (1966).

[23] S. Yajima, M. Hamano, and N. Abe, unpublished work.

[24] C. A. Smith and C. T. Young, NAA-SR-72 (1951).

[25] C. T. Young and C. A. Smith, NAA-SR-232 (1953).

[26] L. B. Doyle, NAA-SR-255 (1953).

[27] G. A. Cowan and C. J. Orth, in Proc. 2nd Intern. Conf. Peaceful Uses Atomic Energy, Geneva, Vol. 7, 1958, p. 328.

[28] C. J. Orth, Nucl. Sci. Eng., 9, 417 (1961).

[29] A. R. Saunders, ORNL-3145 (1961).

[30] N. R. Large and G. N. Walton, AERE C/M 346 (1958).

[31] E. A. Bryant, G. A. Cowan, J. E. Sattizahn, and K. Wolfsberg, Nucl. Sci. Eng., 15, 288 (1963).

[32] S. Yajima, K. Shiba, and M. Handa, Bull. Chem. Soc. Japan, 36, 253 (1963).

[33] S. Yajima, K. Shiba, and M. Handa, Bull. Chem. Soc. Japan, 36, 258 (1963).

[34] S. Yajima, K. Shiba, and M. Handa, Bull. Chem. Soc. Japan, 38, 278 (1965).

[35] M. Handa and K. Shiba, private communication.

[36] B. V. Erhler and F. S. Lapteva, J. Nucl. Energy, II, 4, 471 (1957).

[37] M. D. Rogers and J. Adam, J. Nucl. Mater., 6, 182 (1962).

[38] T. K. Bierlein and B. Mastel, J. Nucl. Mater., 7, 32 (1962).

[39] G. J. Dienes and W. E. Parkins, NAA-SR-60 (1950).

[40] K. Shiba, M. Handa, and S. Yajima, J. Nucl. Sci. Tech., 6, 333 (1969).

[41] K. W. Bagnall, The Chemistry of Selenium, Tellurium and Polonium, Elsevier, Amsterdam, 1966, p. 56.

[42] L. K. Matson, J. W. Moody, and R. C. Himes, J. Inorg. Nucl. Chem., 25, 795 (1963).

[43] W. K. Woods, L. P. Bupp, and J. F. Fletcher, in Proc. 1st Intern. Conf. Peaceful Uses Atomic Energy, Geneva, p/746, 7, 455 (1956).

[44] G. H. Kinchin, in Proc. 1st Intern. Conf. Peaceful Uses Atomic Energy, Geneva, p/442, 7, 472 (1956).

[45] R. J. Harrison, CF-53-3-276 (Del.), Radiation Damage Conference, 1954, Part I, p. 93.

[46] T. Iwata and H. Suzuki, "A Model of Radiation Damage in Graphite," I.A.E.A. Symposium on Radiation Damage in Reactor Materials, Venice, 1962, I.A.E.A., Vienna, 1963.

[47] R. C. Croft, Aust. J. Chem., 9, 184 (1956).

AUTHOR INDEX

Plain numbers indicate pages where a name appears. Numbers in parentheses indicate the reference for all preceding page numbers given. Each reference is in turn followed by an underlined number, which shows where it is listed.

Other books of interest to you...

Because of your interest in our books, we have included the following catalog of books for your convenience.

Any of these books are available on an approval basis. This section has been reprinted in full from our **material science** catalog.

If you wish to receive a complete catalog of MDI books, journals and encyclopedias, please write to us and we will be happy to send you one.

MARCEL DEKKER, INC.
95 Madison Avenue, New York, N.Y. 10016

material science

including
Polymers, Plastics, Fibers, and Coatings
Metals and Metallurgy
Ceramics and Glass
Vacuum Science

ALTGELT and SEGAL
Gel Permeation Chromatography

edited by KLAUS H. ALTGELT, *Chevron Research Company, Richmond, California*, and LEON SEGAL, *South Regional Research Laboratory, U.S.D.A., New Orleans, Louisiana*

672 pages, illustrated. 1971

Demonstrates the manifold applications of gel permeation chromatography in the field of polymer chemistry. Directed to all research, quality-control, and analytical chemists working with conventional and unconventional polymers and other large molecules in the fields of polymer, cellulose, and petroleum chemistry.

CONTENTS: The sizes of polymer molecules and the GPC separation, *F. W. Billmeyer, Jr. and K. H. Altgelt.* Gel permeation chromatography column packings — types and uses, *D. J. Harmon.* Chromatographic instrumentation and detection of gel permeation effluents, *E. M. Barrall, II and J. F. Johnson.* Peak resolution and separation power in gel permeation chromatography, *D. J. Harmon.* A review of peak broadening in gel chromatography, *R. N. Kelley and F. W. Billmeyer, Jr.* Mathematical methods of correcting instrumental spreading in GPC, *L. H. Tung.* Comparison of different techniques of correcting for band broadening in GPC, *J. H. Duerksen.* Separation mechanisms in gel permeation chromatography, *W. W. Yau, C. P. Malone, and H. L. Suchan.* Gel permeation chromatography and thermodynamic equilibrium. *E. F. Casassa.* Calibration of GPC, columns, *H. Coll.* Data treatment in GPC, *L. H. Tung.* The overload effect in gel permeation chromatography, *J. C. Moore.* Gel permeation chromatography using a bio-glas substrate having a broad pore size distribution, *A. R. Cooper, J. H. Cain, E. M. Barrall, II, and J. F. Johnson.* High resolution gel permeation chromatography — using recycle, *K. J. Bombaugh and R. F. Levangie.* Gel permeation chromatography with high loads, *K. H. Altgelt.* Fast gel permeation chromatography, *J. N. Little, J. L. Waters, K. J. Bombaugh, and W. J.*

Pauplis. Extension of GPC techniques, *G. Meyerhoff.* Phase distribution chromatography (PDC) of polystyrene, *R. H. Casper and G. V. Schulz.* Apparent and real distribution in GPC (experiments with PMMA samples), *K. C. Berger and G. V. Schulz.* The instrument spreading correction in GPC. I: The general shape function using a linear calibration curve, *T. Provder and E. M. Rosen.* The instrument spreading correction in GPC. II: The general shape function using the Fourier transform method with a nonlinear calibration curve, *E. M. Rosen and T. Provder.* Behavior of micellar solutions in gel permeation chromatography: A theory based on a simple model, *H. Coll.* Gel permeation analysis of macromolecular association by an equilibrium method, *B. F. Cameron, L. Sklar, V. Greenfield, and A. D. Adler.* Gel filtration chromatography, *B. F. Cameron.* Determination of polymer branching with gel permeation chromatography. Abstract of a review, *E. E. Drott and R. A. Mendelson.* Fractionation of linear polyethylene with gel permeation chromatography. Part III, *N. Nakajima.* Application of GPC in the study of stereospecific block copolymers, *R. D. Mate and M. R. Ambler.* Composition of butadiene-styrene copolymers by gel permeation chromatography, *H. E. Adams.* A direct GPC calibration for low molecular weight polybutadiene, employing dual detectors, *J. R. Runyon.* Quantitative determination of plasticizers in polymeric mixtures by GPC, *D. F. Alliet and J. M. Pacco.* Evaluation of pulps, rayon fibers, and cellulose acetate by GPC and other fractionation methods, *W. J. Alexander and T. E. Muller.* Characterization of the internal pore structures of cotton and chemically modified cottons by gel permeation, *L. F. Martin, F. A. Blouin, and S. P. Rowland.* Application of GPC to studies of the viscose process. I: Evaluation of the method, *L. H. Phifer and J. Dyer.* Application of GPC to studies of the viscose process. II: The effects to steeping and alkali-crumb aging, *J. Dyer and L. H. Phifer.* Gel permeation chromatography calibration. I: Use of calibration curves based on polystyrene in THF and integral distribution curves of elution volume to generate calibration curves for polymers in 2,2,2-trifluoroethanol, *T. Provder, J. C. Woodbrey, and J. H. Clark.* Modification of a gel permeation chromatograph for automatic sam-

(continued)

ALTGELT and SEGAL *(continued)*

ple injection and on-line computer data recording, *A. R. Gregges, B. F. Dowden, E. M. Barral, II, and T. T. Horikawa.* Characterization of crude oils by gel permeation chromatography, *H. H. Oelert, D. R. Latham, and W. E. Haines.* Separation and characterization of high-molecular-weight saturate fractions by gel permeation chromatography, *J. H. Weber and H. H. Oelert.* Fractionation of residuals by gel permeation chromatography, *E. W. Albaugh, P. C. Talarico, B. E. Davis, and R. A. Wirkkala.* Combined gel permeation chromatography–NMR techniques in the characterization of petroleum residuals, *F. E. Dickson, R. A. Wirkkala, and B. E. Davis.* A rapid method of identification and assessment of total crude oils and crude oil fractions by gel permeation chromatography, *J. N. Done and W. K. Reid.* Gel permeation analysis of asphaltenes from steam stimulated oil wells, *C. A. Stout and S. W. Nicksic.* GPC separation and integrated structural analysis of petroleum heavy ends, *K. H. Altgelt and E. Hirsch.*

AMERICAN VACUUM SOCIETY
Experimental Vacuum Science and Technology

edited by THE AMERICAN VACUUM SOCIETY EDUCATION COMMITTEE

288 pages, illustrated. 1973

A collection of experiments, which are graded from simple procedures to sophisticated vacuum processes, and designed to aid instructors and introduce students to the basic concepts and techniques of the field of vacuum science. Includes an extensive bibliography to stimulate further investigation. Especially useful for all students and teachers in the many fields of the basic sciences and engineering where vacuum methods and techniques are important.

CONTENTS: **Section 1: Procedures in Vacuum Production and Measurement,** *W. Brunner and H. Patton.* **Section 2: Experiments which Illustrate the Characteristics of the Vacuum Environment:** Demonstration of the outgassing of different vacuum materials, *J. Rosebury.* Comparison of gas evolution phenomenon from glass and metal system envelopes during baking, *R. Lawson.* Determination of the net quantity of gas flowing through a cylindrical tube, *K. Busen.* **Section 3: Experiments which Illustrate the Dependence of the Physical Properties of Gases on Gas Density:** Measurement of the pumping action of an ionization gauge, *H. Farber.* Study of the linearity of an ionization gauge, *J. Miller, III.* Calibration of gauges, *C. Morrison.* **Section 4: Experiments which Examine Physical and Chemical Interactions at Surfaces:** Study of the sorption of gases for different gas–sorbent combinations, *K. Wear.* The use of sorbents as traps and pumps, *H. Farber.* Sorption of gases

by titanium, *H. Farber.* Investigation of the passage of oxygen across a silver barrier, *K. Busen.* **Section 5: Processes Requiring a Vacuum Environment:** Thin film evaporation, *M. Thomas.* Fabrication of a nichrome resistor, *R. Riegert and G. Breitweiser.* Sputtering, *P. Grosewald.* Ejection patterns in single crystal sputtering, *G. Wehner.* **Section 6: Special Projects:** Study of the sublimation of ice at various pressures, *W. Parker.* Study of friction, *P. McElligott.* Measurement of the mean free path of conduction electrons in silver, *R. Olson and J. Wilson.* Construction and use of a cathode ray tube, *B. Kendall and H. Luther.* Construction of a vacuum triode using solder glass techniques, *J. King and J. Orsula.* Experiments using solder glass techniques, *D. Whitcomb.* **Section 7: Speculations:** Original thought experiments, *M. Carbone.* Provocative ideas and questions, *N. Milleron.*

BEER   *Liquid Metals:* Chemistry and Physics

(Monographs and Textbooks in Material Science Series, Volume 4)

edited by SYLVAN Z. BEER, *Converta Enterprises, Inc., Syracuse, New York*

742 pages, illustrated. 1972

Presents a comprehensive review of the research done on the liquid state of metals, bringing together the latest advances, as well as data previously scattered among a wide variety of publications. Of prime importance to chemists, physicists, research metallurgists, metallurgical engineers, and materials scientists working in the areas of liquid-state theory, the theory of metals, process metallurgy involving liquid metals, and high-temperature chemistry.

CONTENTS: On the thermodynamic formalism of metallic solutions, *C. H. P. Lupis.* Kinetics of evaporation of various elements from liquid iron alloys under vacuum, *R. Ohno.* Relation between thermodynamic and electrical properties of liquid alloys, *D. N. Lee and B. D. Lichter.* The surface tension of liquid metals, *B. C. Allen.* Significant structure theory applied to liquid metals, *S. M. Breitling and H. Eyring.* Diffraction analysis of liquid metals and alloys, *C. N. J. Wagner.* The optical properties of liquid metals, *J. N. Hodgson.* Effect of pressure on the properties of liquid metals, *A. Rapoport.* Sound propagation in liquid metals, *R. T. Beyer and E. M. Ring.* The viscosity of liquid metals, *R. T. Beyer and E. M. Ring.* Magnetic properties of liquid metals, *R. Dupree and E. F. W. Seymour.* Diffusion in liquid metals, *N. H. Nachtrieb.* Electromigration in liquid alloys, *S. G. Epstein.* Electronic nature of liquid metals and liquid metal theory, *J. E. Enderby.* Structure and properties of noncrystalline metallic alloys produced by rapid quenching of liquid alloys, *B. C. Giessen and C. N. J. Wagner.*

BLACK and PRESTON *High-Modulus Wholly Aromatic Fibers*

(Fiber Science Series, Volume 5)

edited by W. BRUCE BLACK, *Monsanto Textiles Company, Pensacola, Florida* and JACK PRESTON, *Monsanto Textiles Company, Chemstrand Research Center, Durham, North Carolina*

304 pages, illustrated. 1973

Based on a symposium on high-modulus aromatic fibers held by the American Chemical Society in Boston on April 13, 1972. The first formal publication of research which shows the relationship of fiber properties to polymer structure. Extremely significant reading for all fiber scientists and material scientists; plastics scientists and engineers interested in fiber-reinforced plastics; spacecraft and aircraft oriented engineers; and scientists in the industrial fiber, sports equipment, and airframe fields.

CONTENTS: High-modulus wholly aromatic fibers: Introduction to the Symposium and historical perspective, *W. Black.* High-modulus wholly aromatic fibers. I. Wholly ordered poly-amide-hydrazines and poly-1,3,4,-oxadizole-amides, *J. Preston, W. Black and W. Hofferbert, Jr.* High-modulus wholly aromatic fibers. II. Partially ordered polyamide-hydrazides, *J. Preston, W. Black, and W. Hofferbert, Jr.* Self-regulating polycondensations. II. A study of the order present in polyamide-hydrazides derived from terephthaloyl chloride and p-aminobenz-hydrazide, *R. Morrison, J. Preston, J. Randall, and W. Black.* Self-regulating polycondensations. III. NMR analysis of oligomers derived from terephthaloyl chloride and p-aminobenz-hydrazide, *J. Randall, R. Morrison, and J. Preston.* Some physical and mechanical properties of some high-modulus fibers prepared from all-para aromatic polyamide-hydrazides, *W. Black, J. Preston, H. Morgan, G. Raumann, and M. Lilyquist.* Morphology and crystal structure of wholly aromatic all-para polyamide-hydrazide polymers, *V. Holland.* X-ray study of an all-para wholly aromatic polyamide-hydrazide[a,b], *R. Miller.* Molecular weight characterization of wholly para-oriented, aromatic polyamide-hydrazides and wholly aromatic polyamides, *J. Burke.* Construction and properties of fabrics of high-modulus organic fibers useful for composite reinforcing, *M. Lilyquist, R. DeBrunner, and J. Fincke.* Mechanical properties of a high-modulus polyamide-hydrazide fiber in composites and of the polyamide-hydrazide fiber and fabric composites, *D. Zaukelies and B. Daniels.* Tire cord application of high-modulus fibers derived from polyamide-hydra-zides, *G. Raumann and J. Brownlee.* The application of high-modulus fibers to ballistic protection, *R. Laible, F. Figucia, and W. Ferguson.* High-modulus wholly aromatic fibers. III. Random copolymers containing hydrazide and/or carbonamide linkages, *J. Preston, H. Morgan, and W. Black.*

BOLKER *Natural and Synthetic Polymers: An Introduction*

by HENRY I. BOLKER, *Department of Chemistry, McGill University, Montreal, Quebec*

in preparation. 1973

Presents a unified approach to polymer chemistry, with equal emphasis on natural and synthetic polymers, and is arranged in a logical sequence of topics based on increasing complexity of molecular architecture. Useful as a textbook for a first course in polymer chemistry and as a reference book for workers in the field.

CONTENTS: Introduction • Natural condensation polymers: The linear polysaccharides • Synthetic condensation (step-growth) polymers • Addition (chain-growth) polymers • Stereoregularity in addition polymers • Branched homopolymers: Synthetic and natural • Natural heteropolymers: I. Heteropolysaccharides • Natural heteropolymers: II. Nucleic acids • Copolymers and copolymerization • Crosslinking in synthetic polymers • Natural heteropolymers: III. Polypeptides and proteins • Lignins.

BROWNING *Analysis of Paper*

by B. L. BROWNING, *The Institute of Paper Chemistry, Appleton, Wisconsin*

352 pages, illustrated. 1969

Provides comprehensive coverage of methods for chemical analysis of paper. Is of value to manufacturers of paper and paper board, suppliers of components or of additives introduced into paper, converters and printers, purchasers and users, librarians, and others concerned with the properties, behavior, and applications of paper that are related to composition.

CONTENTS: Paper as a commodity • Sampling and preparation of sample • Determination of moisture • Fiber analysis • Fiber quality methods • Lignin • Rosin size • Starch • Proteins • Coatings • Waxes and oils • Fillers and white coating pigments • Dyes and colored pigments • Acidity and alkalinity • Residues and impurities • Biological control agents • General identification of additives in paper • Synthetic resins • Wet-strength agents • Polysaccharides and gums • Miscellaneous additives • Noncellulose fibers • Specks and spots • Permanence of paper • Paper in forensic science.

BUTLER, O'DRISCOLL, and SHEN *Reviews in Macromolecular Chemistry*

(Book Edition)

edited by GEORGE B. BUTLER, *Department of Chemistry, University of Florida,*

(continued)

BUTLER, O'DRISCOLL, and SHEN *(continued)*

Gainesville, and KENNETH F. O'DRISCOLL, *Department of Chemical Engineering, University of Waterloo, Ontario, Canada* and MITCHEL SHEN, *Department of Chemical Engineering, University of California, Berkeley*

Vol. 1 *out of print*

Vol. 2 388 pages, illustrated. 1968

Vol. 3 430 pages, illustrated. 1969

Vol. 4 428 pages, illustrated. 1970

Vol. 5, Part I see NEUSE and ROSENBERG

Vol. 5, Part II
250 pages, illustrated. 1970

Vol. 6 498 pages, illustrated. 1971

Vol. 7 314 pages, illustrated. 1972

Vol. 8 346 pages, illustrated. 1972

Vol. 9 380 pages, illustrated. 1973

Reviews of the currently published literature for those who wish to keep abreast of the new and rapidly advancing developments in macromolecular chemistry. Of interest to organic and physical chemists, biochemists, engineers, and all students and research workers in polymer chemistry and related fields.

CONTENTS:

Volume 1: Application of molecular orbital theory to vinyl polymerization, *K. F. O'Driscoll and T. Yonezawa.* Poly(alkylene oxides), *A. E. Gurgiolo.* Polyurethanes, *D. J. Lyman.* Uncatalyzed, uninhibited thermal oxidation of saturated polyolefins, *L. Reich and S. S. Stivala.* Double-strand polymers, *W. De Winter.* Biomedical polymers, *D. J. Lyman.* Gel permeation chromatography with organic solvents, *J. F. Johnson, R. S. Porter, and M. J. R. Cantow.*

Volume 2: Phosphorus-containing polymers: Introduction, *M. Sander and E. Steininger.* Linear polymers with phosphorus in side chains, *M. Sander and E. Steininger.* Linear polymers with phosphorus and carbon in the main chain, *M. Sander and E. Steininger.* Reassessment of the theory of polyesterification with particular reference to alkyd resins, *D. H. Solomon.* Symmetry considerations for stereoregular polymers, *A. M. Liquori.* Copolymerization of vinyl monomers with ring compounds, *R. A. Patsiga.* Application of high-resolution nuclear magnetic resonance to polymer structure determination, I., *K. C. Ramey and W. S. Brey, Jr.* Ten years of polymer single crystals, *D. A. Blackadder.* Thermal degradation of polystyrene, *G. G. Cameron and J. R. MacCallum.*

Volume 3: Phosphorous-containing resins, *M. Sander and E. Steininger.* Inorganic phosphorous polymers, *M. Sander and E. Steininger.* Phosphorylation of polymers, *M. Sander and E. Steininger.* Sulphur-containing polymers, *E. J. Goethals.* Polymer molecular weight distributions, *N. Amundson and D. Luss.* Heteroatom

ring-containing polymers, *A. D. Delman.* Molecular theories of rubber-like elasticity and polymer viscoelasticity, *M. Shen, W. F. Hall, and R. E. DeWames.* End-group studies using dye techniques, *S. R. Palit and B. M. Mandal.* Free-radical spin labels for macromolecules, *J. D. Ingham.* The synthesis of thermally stable polymers: A progress report, *J. I. Jones.*

Volume 4: Polymer enzymes and enzyme analogs, *A. S. Lindsey.* Stability of polycarbonate, *A. Davis and J. Golden.* Cross-linking — effect on physical properties of polymers, *L. E. Nielsen.* The synthesis of thermally stable polymeric azomethines by polycondensation reactions, *G. F. D'Alelio and R. K. Schoenig.* On the dehydrochlorination and the stabilization of polyvinyl chloride, *M. Onozuka and M. Asahina.* Recent advances in the development of flame-retardant polymers, *A. D. Delman.* Thermodynamics of polymerization. I, *H. Sawada.*

Vol. 5, Part II: Ring-chain equilibria, *H. Allcock.* Occupied volume of liquids and polymers, *R. Haward.* The application of ESR techniques to high polymer fracture, *H. Kausch-Blecken von Schmeling.* The science of determining copolymerization reactivity ratios, *P. Tidwell and G. Mortimer.* Block polymers and related heterophase elastomers, *G. Estes, S. Cooper, and A. Tobolsky.*

Volume 6: Proton magnetic resonance of molecular interactions in polymer solutions, *K.-J. Liu and J. E. Anderson.* Preparation and polymerization of vinyl heterocyclic compounds, *K. Takemoto.* Catalysis in isocyanate reactions, *K. C. Frisch and L. P. Rumao.* Thermodynamics of polymerization. II. Thermodynamics of ring—opening polymerization, *H. Sawada.* Copolymers of naturally occurring macromolecules, I. *C. Watt.* Molecular configuration and pyrolysis of phenolic-novolaks, *E. L. Winkler and J. A. Parker.* Physical properties of ionic polymers, *E. P. Otocka.* Synthesis and properties of polyphenyls and polyphenylenes, *J. G. Speight, P. Kovacic, and F. W. Koch.* Dependence of flow properties on molecular weight, temperature, and shear, *A. Casale, R. S. Porter, and J. F. Johnson.* Synthesis methods and properties of polyazoles, *V. V. Korshak and M. M. Teplyakov.*

Volume 7: Linear polyquinoxalines, *P. M. Hergenrother.* Nylons—known and unknown. A comprehensive index of linear aliphatic polyamides of regular structure, *H. K. Livingston, M. S. Sioshansi, and M. D. Glick.* Recent advances in polymer fractionation, *L. H. Tung.* Rheology of adhesion, *D. H. Kaelble.* Solvation of synthetic and natural polyelectrolytes, *B. E. Conway.* Hydrogen transfer polymerization with anionic catalysts and the problem of anionic isomerization polymerization, *J. P. Kennedy and T. Otsu.*

Volume 8: Polymerization by carbenoids, carbenes, and nitrenes, *M. Imoto and T. Nakaya.* Collagen and gelatin in the solid state, *I. V. Yannas.* Ring-opening polymerization of cycloolefins, *N. Calderon.* Thermodynamics of polymerization. III, *H. Sawada.* Polymerization of N-vinylcarbazole initiated by metal salts, *M. Biswas and D. Chakravarty.* Vibrational spectroscopy of polymers, *F. J. Boerio and J. L. Koenig.* Polymer compatibility, *S. Krause.*

Volume 9: Mechanical properties of polymers: The influence of molecular weight and molecular weight distribution, *J. Martin, J. Johnson, and A. Cooper.* On the mathematical modeling of polymerization reactors, *W. Ray.* Anionic cyclopolymerization, *C. McCormick and G. Butler.* Carbon-13 NMR of polymers, *V. Mochel.* Thermodynamics of polymerization. IV. Thermodynamics of equilibrium polymerization, *H. Sawada.*

CARROLL *Physical Methods in Macromolecular Chemistry*

a series edited by BENJAMIN CARROLL, *Rutgers—The State University, Newark, New Jersey*

Vol. 1 400 pages, illustrated. 1969

Vol. 2 384 pages, illustrated. 1972

A series which reviews why and how analytical methods are used in the study of macromolecules. Each method is critically discussed by experts in the field. Directed to researchers in polymer chemistry, biopolymers, and organic and inorganic chemistry.

CONTENTS:

Volume 1: Surface chemistry and polymers, *M. Rosoff.* Internal reflection spectroscopy, *J. K. Barr and P. A. Flournoy.* Electric properties of synthetic polymers, *E. O. Forster.* Assessing radiation effects in polymers, *P. Y. Feng and E. S. Freeman.* Fluorescence techniques for polymer solutions, *D. J. R. Laurence.* Insoluble polymers: Molecular weights and their distributions, *H. C. Cheung.*

Volume 2: Gel permeation chromatography in polymer chemistry, *D. D. Bly.* Interactions of polymers with small ions and molecules, *D. J. R. Laurence.* Electric properties of biopolymers: Proteins, *E. O. Forster and A. P. Minton.* Thermal methods, *E. P. Manche and B. Carroll.*

CARTER *Essential Fiber Chemistry*

(Fiber Science Series, Volume 2)

by MARY E. CARTER, *FMC Corporation, American Viscose Division, Marcus Hook, Pennsylvania*

232 pages, illustrated. 1971

Discusses the chemical and physical structure and properties of ten commercially important fibers. Useful to all chemists interested in the research and development of natural and man-made fibers.

CONTENTS: Cotton • Rayon • Cellulose acetate • Wool • Polyamide • Acrylic fibers • Polyethylene terephthalate • Polyolefins • Spandex • Glass.

CONLEY *Thermal Stability of Polymers*

In 2 Volumes

(Monographs in Macromolecular Chemistry Series)

edited by R. T. CONLEY, *Wright State University, Dayton, Ohio*

Vol. 1 656 pages, illustrated. 1970

Vol. 2 in preparation. 1974

CONTENTS: Introduction, *R. T. Conley.* Molecular structure and stability criteria, *R. T. Conley.* The relationship between the kinetics and mechanism of thermal depolymerization, *R. H. Boyd.* Random scission processes, *A. V. Tobolsky, A. M. Kotliar, and T. C. P. Lee.* Fundamental reactions in oxidation chemistry, *P. M. Norling and A. V. Tobolsky.* Thermal and oxidative degradation of polyethylene, polypropylene, and related olefin polymers, *R. H. Hansen.* Thermal and oxidative degradation of natural rubber and allied substances, *E. M. Bevilacqua.* Vinyl and vinylidene polymers, *R. T. Conley and R. Malloy.* Fluorocarbon polymers, *W. W. Wright.* Thermal and thermooxidative degradation of polyamides, polyesters, polyethers, and related polymers, *R. T. Conley and R. A. Gaudiana.* Thermosetting resins, *R. T. Conley.* Thermal and thermooxidative degradation of cellulosic polymers, *R. T. Conley.* Heterocyclic polymers, *G. P. Shulman.* Degradation of inorganic polymers, *J. Economy and J. H. Mason.*

D'ALELIO and PARKER

Ablative Plastics

edited by GAETANO F. D'ALELIO, *Department of Chemistry, University of Notre Dame, Indiana,* and JOHN A. PARKER, *NASA, Ames Research Center, Moffet Field, California*

504 pages, illustrated. 1971

Provides the comprehensive and rational approach required for the design and production of reliable head shields for future space missions. Includes discussions on the various aspects of heat-rejection mode as a function of heating rate; the nature of the heat transfer, both radiative and conductive; and the nature of degrading polymers. A valuable reference for all aerospace scientists, polymer chemists, physicists, and aerodynamic engineers.

CONTENTS: Ablative polymers in aerospace technology, *D. L. Schmidt.* Hypervelocity heat protection—a review of laboratory experiments, *N. S. Vojvodich.* A review of ablative studies of interest to naval applications, *F. J. Koubek.* Structural design and thermal properties of polymers, *G. F. D'Alelio.* Characterization of an epoxy-anhydride ablative system using com-

(continued)

D'ALELIO and PARKER *(continued)*

puter treatment of analytical results, *C. G. Taylor and E. L. Pendleton.* The synthesis and characterization of some potential ablative polymers, *R. Y. Wen, L. F. Sonnabend, and R. Eddy.* Thermal degradation and curing of polyphenylene, *D. N. Vincent.* Thermosetting polyphenylene resin—its synthesis and use in ablative composites, *N. Bilow and L. J. Miller.* Structural ablative plastics, *R. M. Lurie, S. F. D'Urso, and C. K. Mullen.* Prediction of heat shield performance in terms of epoxy resin structure, *G. J. Fleming.* Ablative resins for hyperthermal environments, *B. S. Marks and L. Rubin.* The development of polybenzimidazole composites as ablative heat shields, *R. R. Dickey, J. H. Lundell, and J. A. Parker.* Ablative degradation of a silicon foam, *T. McKeon.* Thermophysical characteristics of high-performance ablative composites, *M. L. Minges.* The design and development of a high-heating rate thermogravimetric analyzer suitable for use with ablative plastics, *A. M. Melnick and E. J. Nolan.* Pyrolysis kinetics of nylon 6-6, phenolic resin and their composites, *H. E. Goldstein.* Pyrolysis-gas chromatography as a tool for studying the degradation of ablative plastics, *R. M. Ross.* Nonequilibrium flow and the kinetics of chemical reactions in the char zone, *G. C. April, R. W. Pike, and E. G. del Valle.* Arc-image testing of ablation materials, *E. M. Liston.* Development and characterization of a radio frequency-transparent ablator, *E. L. Strauss.* Tailoring polymers for entry into the atmosphere of Mars and Venus, *R. G. Nagler.*

DIGGLE   *Oxides and Oxide Films*

in multi-volumes

(The Anodic Behavior of Metals and Semiconductors Series)

edited by JOHN W. DIGGLE, *Research School of Chemistry, The Australian National University, Canberra*

Vol. 1   552 pages, illustrated. 1972

Vol. 2   424 pages, illustrated. 1973

Treats the anodic behavior of metals and semiconductors and peripheral areas in an authoritative and interdisciplinary manner. The initial volumes deal with the physics and chemistry of oxides and oxide films. Of great value for all those involved in electrochemistry, materials science, solid state physics, electrical engineering, metallurgy, corrosion science, semiconductors, and electrochemical technology.

CONTENTS:

Volume 1: Passivation and passivity, *V. Brusić.* Mechanisms of ionic transport through oxide films, *M. J. Dignam.* Electronic current flow through ideal dielectric films, *C. A. Mead.* Electrical double layer at metal oxide-solution interfaces, *S. Ahmed.*

Volume 2: Anodic oxide films: Influence of solid-state properties on electrochemical behavior, *A. Vijh.* Dielectric loss mechanism in amorphous oxide films, *D. M. Smyth.* Porous anodic films in aluminum, *G. C. Wood.* Dissolution of oxide phases, *J. W. Diggle.*

FOURT and HOLLIES
Clothing: Comfort and Function

(Fiber Science Series, Volume 1)

by LYMAN FOURT and NORMAN HOLLIES, *Gillette Research Institute, Rockville, Maryland*

272 pages, illustrated. 1970

A unified review of the present state of knowledge in the science of clothing. Of interest to textile scientists, fiber producers and marketers, textile converters, and garment makers.

CONTENTS: The factors involved in the study of clothing • Clothing considered as a system interacting with the body • Clothing considered as a structural assemblage of materials • Heat and moisture relations in clothing • Physiological and field testing of clothing by wearing it • Physical properties of clothing and clothing materials in relation to comfort • Differences between fibers with respect to comfort • Current trends and new developments in the study of clothing.

FRISCH and REEGEN   *Ring-Opening Polymerization*

(Kinetics and Mechanisms of Polymerization Series, Volume 2)

edited by KURT C. FRISCH, and SIDNEY L. REEGEN, *Polymer Institute, University of Detroit, Michigan*

544 pages, illustrated. 1969

Covers the polymerization of important classes of cyclic monomers such as ethylene and propylene oxide, alkylenimines, and sulfides, lactones, lactams, cyclic silicone compounds, and cyclic nitrogen containing heterocycles. Of great interest to the industrial, commercial, and academic worlds as it has application to elastomers, coatings, fibers, films, and foams.

CONTENTS: 1,2 Epoxides, *Y. Ishii and S. Sakai.* 1,3 Epoxides and higher epoxides, *P. Dreyfuss and M. P. Dreyfuss.* Cyclic formals, *J. Furukawa and K. Tada.* Cyclic sulfides, *P. Sigwalt.* Alkylenimines, *M. Hauser.* Lactones, *R. D. Lundberg and E. F. Cox.* Lactams, *H. K. Reimschuessel.* Cyclic siloxanes and silazanes, *E. E. Bostick.* Nitrogen-containing heterocyclic compounds, *V. Kargin and V. Kabanov.* N-carboxy-α-amino acid anhydrides, *Y. Shalitin.*

FRISCH and SAUNDERS
Plastic Foams

(Monographs on Plastic Series, Volume 1)

edited by KURT C. FRISCH, *University of Detroit, Michigan*, and JAMES H. SAUNDERS, *Monsanto Company, Pensacola, Florida*

Part I 464 pages, illustrated. 1972

Part II 704 pages, illustrated. 1973

Gives an integrated picture of the fundamental principles, technology, and applications of foams, and offers a thorough treatment of specific types of plastic foams. Emphasis is placed on the newer trends in this science.

Of particular value to chemists and engineers engaged in research and development, and marketing and production personnel in the polymer and plastics industry.

CONTENTS:

Part I: Introduction, *K. C. Frisch.* The mechanism of foam formation, *J. H. Saunders and R. H. Hansen.* Flexible polyurethane foams, *G. T. Gmitter, H. J. Fabris, and E. M. Maxey.* Sponge rubber and latex foam, *R. L. Zimmerman and H. R. Bailey.* Polyolefin foams, *D. J. Sundquist.* Polyvinyl chloride foams, *A. C. Werner.* Silicone foams, *H. L. Vincent.* Testing of cellular materials, *R. A. Stengard.*

Part II: Rigid urethane foams, *J. K. Backus and P. G. Gemeinhardt.* Polystyrene and related thermoplastic foams, *A. R. Ingram and J. Fogel.* Phenolic foams, *A. J. Papa and W. R. Proops.* Urea-formaldehyde foams, *K. C. Frisch.* Epoxy-resin foams, *H. Lee and K. Neville.* New high-temperature-resistant plastic foams, *E. E. Hardy and J. H. Saunders.* Miscellaneous foams, *K. C. Frisch.* Inorganic foams, *M. Wismer.* Effects of cell geometry on foam performance, *R. H. Harding.* Thermal decomposition and flammability of foams, *P. E. Burgess, Jr. and C. J. Hilado.* Foams in transportation, *M. Kaplan and L. M. Zwolinski.* Architectural uses of foam plastics, *S. C. A. Paraskevopoulos.* Military and space applications of cellular materials, *R. J. F. Palchak.*

GARG, SVALBONAS, and GURTMAN
Analysis of Structural Composite Materials

(Monographs and Textbooks in Material Science Series, Volume 6)

by SABODH GARG, *Systems, Science and Software Company, La Jolla, California*, VYTAS SVALBONAS, *Grumman Aerospace Corporation, Bethpage, New York*, and GERRY GURTMAN, *Systems, Science and Software Company, La Jolla, California*

552 pages, illustrated. 1973

Compares various theories for the static and dynamic analysis of structural composite materials. Deals with the elastic properties of laminated composites and particulate and unidirectional fiber reinforced composites, composite strength, and stress wave propagation. May be used as a textbook for a graduate course in composites and is of interest to researchers and analysts in any industry that uses composites.

CONTENTS: Why composites? • Simple analytic models • Elasticity analyses • Bounds on elastic properties by energy methods • Multilayer laminates • Non-statistical models of composite strength • Statistical tensile strength of fiber and fiber bundles • Composite tensile-strength models • Cumulative weakening model including stress concentrations • Compressive strength of composites • Theory of breaking kinetics • Introduction to elastic wave propagation in composites • Approximate analysis techniques for stress wave propagation in composites • Application of continuum mixture theories to the study of elastic wave propagation in composite materials • Shock waves in composite materials.

HAM   Vinyl Polymerization
In 2 Parts

(Kinetics and Mechanisms of Polymerization Series, Volume 1)

edited by GEORGE E. HAM, *Geigy Chemical Corporation, Ardsley, New York*

Part I 560 pages, illustrated. 1967

Part II 432 pages, illustrated. 1969

"The book is a good introduction to the series. It has provided a sound basis for subsequent volumes and should serve as an important reference text for students and researchers in polymer chemistry."—B. D. Gesner, Bell Telephone Labs., *SPE Journal*

"The book is highly recommended."—Arthur Tobolsky, *The American Scientist*

CONTENTS:

Part I: General aspects of free-radical polymerization, *G. E. Ham.* The mechanism of cyclopolymerization of nonconjugated diolefins, *W. E. Gibbs and J. M. Barton.* Styrene, *M. H. George.* Mechanism of vinyl acetate polymerization, *M. K. Lindemann.* Polymerization of vinyl chloride and vinylidene chloride, *G. Talamini and E. Peggion.* Occlusion phenomena in the polymerization of acrylonitrile and other monomers, *A. D. Jenkins.* Polymerization of acrolein, *R. C. Schulz.* Heats of polymerization and their structural and mechanistic implications, *R. M. Joshi and B. J. Zwolinski.*

Part II: Mechanism of emulsion polymerization, *J. W. Vanderhoff.* Elucidation of emulsion polymerization mechanism based upon copolymer studies, *W. F. Fowler, Jr.* Mechanism of the emulsion polymerization of ethylene, *H. K. Stryker, G. J. Mantell, and A. F. Helin.* Mechanism of stereospecific polymerization of propylene, *W. "E" Smith.* Anionic polymerization,

(continued)

material science

HAM *(continued)*

M. Morton. Mechanisms of cationic polymerization, *Z. Zlámal.* Radiation-induced polymerization, *Y. Tabata.*

HENCH and DOVE *Physics of Electronic Ceramics*

In 2 Parts

(Ceramics and Glass: Science and Technology Series, Volume 2)

edited by LARRY L. HENCH and DEREK B. DOVE, *College of Engineering, University of Florida, Gainesville.*

Part A 584 pages, illustrated. 1971

Part B 576 pages, illustrated. 1972

A highly useful treatise which deals with the physical basis for the behavior of electronic ceramics.

Fundamental physical theories describing each type of electronic ceramics are presented, with discussions included that relate the theories to the applications of the materials. Of special value to graduate students who have had a course in modern physics and also of interest to materials scientists and engineers in electronics, communications, and ceramics.

CONTENTS:

Part A: Quantum mechanics and ceramics, *J. C. Slater.* Band structure and electronic properties of ceramic crystals, *D. Adler.* Electrical conduction in low mobility materials, *I. Bransky and N. M. Tallan.* Defect structure and electronic properties of ceramics, *R. W. Vest.* Conduction domains in solid mixed conductors and electrolytic domain of calcia stabilized zirconia, *J. Patterson.* Semiconducting glasses, *J. D. Mackenzie.* Electronic processes in amorphous semiconductors, *E. A. Davis.* Heterogeneous semiconducting glasses, *H. F. Schaake.* The determination of local order in amorphous semiconducting films, *D. B. Dove.* Negative capacitance effects in amorphous semiconductors, *M. Allen, P. Walsh, and W. Doremus.* Some conduction phenomena in amorphous materials, *K. L. Chopra.* Applications of thin film dielectrics in microelectronics, *N. N. Axelrod.* Substructure and electrical conduction in amorphous thin films, *N. Fuschillo and A. D. McMaster.* Structure of surface defects, *D. L. Stoltz and J. J. Hren.* Electronic surface properties, *P. Mark.* Electron spin resonance and defects in solids, *W. S. Brey, R. B. Gammage, and Y. P. Virmani.* Theory of linear dielectrics, *A. D. Franklin.* Polycrystalline insulators, *H. C. Graham and N. M. Tallan.* Electrical conduction in glass and glass-ceramics, *D. L. Kinser.* Dielectric breakdown of ceramics, *G. C. Walther and L. L. Hench.*

Part B: Some structural mechanisms in ferroelectricity, *R. Pepinsky.* Thermodynamic phenomenology of ferroelectricity in single crystal

and ceramic systems, *L. E. Cross.* Dynamical effects in solid state phase transformations, *J. D. Axe.* Theory of antiferromagnetism and ferrimagnetism, *J. B. Goodenough.* Microstructure and processing of ferrites, *F. J. Schnettler.* Microwave garnet compounds, *G. R. Harrison and L. R. Hodges, Jr.* The optical absorption of glasses, *N. J. Kreidl.* Light scattering from glass, *J. J. Hammel.* Electro-optical and magneto-optical effects, *Y. R. Shen.* The influence of the composition of the gain of Nd-doped glasses, *C. F. Rapp.* Solid state reactions in the preparation of zircon stains, *R. A. Eppler.* Computer color control for ceramic wall tile, *W. K. Culbreth, Jr.* Ceramics and glasses — some uses in the communications industry, *D. G. Thomas.*

HENCH and GOULD *Characterization of Ceramics*

(Ceramics and Glass: Science and Technology Series, Volume 3)

edited by LARRY L. HENCH and ROBERT W. GOULD, *University of Florida, Gainesville*

672 pages, illustrated. 1971

Focuses on the two major directions which comprise the distinct discipline of ceramic characterization: the exploration of the factors that control the properties of the final product and the rapid development of high resolution analytical techniues used for ceramic materials. A particularly timely textbook for an advanced undergraduate or graduate materials science curriculum. Valuable for all materials scientists and ceramic and materials engineers working on the development of an effective and economical materials characterization program.

CONTENTS: Introduction to the characterization of ceramics, *L. L. Hench.* **Part 1: Chemical Analysis:** General analytical chemistry, *P. Rankin.* X-ray spectroscopy, *R. W. Gould.* Atomic absorption flame spectrometry, *J. D. Wineford-ner.* **Part 2: Phase State and Structure:** X-ray diffraction, *R. W. Gould.* Transmission electron microscopy and electron diffraction, *C. F. Tufts.* Analysis of microstructural defects, *R. W. Newman.* Petrographic analysis, *V. D. Fréchette.* Thermal analysis, *R. K. Ware.* Point defect analysis, *W. J. James and G. Lewis.* **Part 3: Size, Shape, Strain, and Surface of Powders:** Physical characterization, *D. R. Lankard and D. E. Niesz.* Small angle x-ray scattering, *R. W. Gould.* X-ray line profile analysis, *R. W. Gould.* Scanning electron microscopy, *S. R. Bates.* Light scattering, *J. H. Boughton.* Characterization of powder surfaces, *L. L. Hench.* **Part 4: Microstructure:** Electron microprobe, *G. Lewis.* Quantitative stereology, *R. T. DeHoff.* Applied stereology, *S. W. Freiman.* **Part 5: Surfaces:** Characterization of ceramic surfaces, *L. Berrin and R. C. Sundahl.*

KATON *Organic Semiconducting Polymers*

(Monographs in Macromolecular Chemistry Series)

edited by J. E. Katon, *Miami University, Oxford, Ohio*

328 pages, illustrated. 1968

CONTENTS: Basic physics of semiconductors, *D. E. Hill*. Theoretical aspects of the electronic behavior of organic macromolecular solids, *H. A. Pohl*. Recent experimental aspects of the electronic behavior of organic macromolecular solids, *S. Kanda and H. A. Pohl*. Semiconducting organic polymers containing metal groups, *B. A. Bolto*. Semiconducting biological polymers, *D. D. Eley*.

KETLEY *The Stereochemistry of Macromolecules*

In 3 Volumes

edited by A. D. Ketley, *W. R. Grace & Co., Clarksville, Maryland*

Vol. 1 424 pages, illustrated. 1967
Vol. 2 400 pages, illustrated. 1967
Vol. 3 476 pages, illustrated. 1968

CONTENTS:

Volume 1: Ziegler-Natta polymerization: Catalysts, monomers, and polymerization procedures, *D. O. Jordan*. The mechanism of Ziegler-Natta catalysis. I. Experimental foundations, *D. F. Hoeg*. Mechanism of Ziegler-Natta polymerization. II. Quantum-chemical and crystal-chemical aspects, *P. Cossee*. Copolymerization of olefins by Ziegler-Natta catalysts, *I. Pasquon, A. Valvassori, and G. Sartori*. Polymerization of dienes by Ziegler-Natta catalysts, *W. Marconi*. Manufacture and commercial applications of stereoregular polymers, *M. Compostella*.

Volume 2: Stereospecific polymerization of vinyl-type monomers and dienes by alkali-metal-based catalysts, *D. Braun*. Stereospecific polymerization of vinyl ethers, *A. D. Ketley*. Ionic polymerization of aldehydes, ketones, and ketenes, *G. F. Pregaglia and M. Binaghi*. Stereospecific polymerization of epoxides, *T. Tsuruta*. Stereochemistry of free-radical polymerizations, *W. Cooper*. Conformational effects induced in polymers by rigid matrices, *N. Marans*. Simple stereoregular polymers in biological systems, *J. N. Baptist*.

Volume 3: Chain conformation and crystallinity, *P. Corradini*. High-resolution nuclear magnetic resonance of synthetic polymers, *J. C. Woodbrey*. Vibrational analyses of the infrared spectra of stereoregular polypropylene, *T. Miyazawa*. Optically active stereoregular polymers, *M. Farina and G. Bressan*. Physical properties of stereoregular polymers in solid state, *J. F. Johnson and R. S. Porter*. Properties of synthetic linear stereoregular polymers in solution, *V. Crescenzi*. Macromolecules as information storage systems, *A. M. Liquori*. Automata theo-

ries of hereditary tactic copolymerization, *H. H. Pattee*. Effect of microtacticity on reactions of polymers, *M. M. van Beylen*. Degradation of stereoregular polymers, *H. H. G. Jellinek*.

KURYLA and PAPA *Flame Retardancy of Polymeric Materials*

a series edited by William C. Kuryla and Anthony J. Papa, *Union Carbide Corporation, South Charleston, West Virginia*

Vol. 1 352 pages, illustrated. 1973
Vol. 2 296 pages, illustrated. 1973

A series concerned with the various modes of rendering polymeric materials fire resistant, which emphasizes specific reagents and techniques in use today. Treats each class of polymer separately to aid the fabricator in gaining an understanding of the specific problems associated with its flammability characteristics, and to review the science and practical solution to its flame retardancy. Of special benefit to industrial polymer chemists, plastics engineers, and fabricators of polymeric materials.

CONTENTS:

Volume 1: Available flame retardants, *W. Kuryla*. Inorganic flame retardants and their mode of action, *J. Pitts*. Fire retardation of polyvinyl chloride and related polymers, *M. O'Mara, W. Ward, D. Knechtges, and R. Meyer*. Fire retardation of wool, nylon, and other natural and synthetic polyamides, *G. Crawshaw, A. Delman, and P. Mehta*.

Volume 2: Fire retardation of polystyrene and related thermoplastics, *R. Lindemann*. Fire retardation of polyethylene and polypropylene, *R. Schwarz*. Flame retardation of natural and synthetic rubbers, *H. Fabris and J. Sommer*. Flame retardancy of phenolic resins and urea- and melamine-formaldehyde resins, *N. Sunshine*.

LEFEVER *Aspects of Crystal Growth*

(Preparation and Properties of Solid State Materials Series, Volume 1)

edited by R. A. Lefever, *Sandia Laboratories, Albuquerque, New Mexico*

Vol. 1 296 pages, illustrated. 1971

Concerns certain aspects of the growth and properties of single crystals. Directed to both beginning and experienced crystal growers, material scientists, and solid state physicists.

CONTENTS: A review of the preparation of single crystals by fused melt electrolysis and some general properties, *W. Kunnmann*. The role of mass transfer in crystallization processes, *W. Wilcox*. Exploratory flux crystal growth, *A. Chase*.

LOWRY *Markov Chains and Monte Carlo Calculations in Polymer Sciences*

(Monographs in Macromolecular Chemistry Series)

edited by GEORGE G. LOWRY, *Western Michigan University, Kalamazoo*

344 pages, illustrated. 1970

Written for the polymer chemist who, although not primarily concerned with mathematical theories, desires a working knowledge of the topics treated. Begins with an introduction to the principles involved and later exemplifies some significant applications of Markov chain theory and Monte Carlo methods.

CONTENTS: Introduction: Deterministic and stochastic approaches, *G. G. Lowry.* Markov chains, *J. Myhre.* Monte Carlo methods, *M. Fluendy.* Polymer conformation as a Markov chain problem, *J. Kinsinger.* Polymer conformation and the excluded-volume problem, *S. Windwer.* Higher order Markov chains and statistical thermodynamics of linear polymers, *J. Mazur.* Copolymer composition and tacticity, *F. P. Price.* Molecular-weight distributions, *G. G. Lowry.*

McCULLOUGH *Concepts of Fiber-Resin Composites*

(Monographs and Textbooks in Material Science Series, Volume 2)

by R. L. McCULLOUGH, *Boeing Scientific Research Laboratories, Seattle, Washington*

128 pages, illustrated. 1971

Presents basic concepts of composite material systems. Introduces the study of composite systems by discussing where and how composite materials are used and how their components are selected. A valuable reference for materials scientists, and students and research management engaged in the exploration of composite materials.

CONTENTS: Materials • Composite structures • Composite properties • The interphase region • Synopsis.

MAY and TANAKA *Epoxy Resins: Chemistry and Technology*

edited by CLAYTON MAY, *Shell Development Company, Emeryville, California* and YOSHIO TANAKA, *Research Institute for Polymers and Textiles, Yokohama, Japan*

704 pages, illustrated. 1973

Brings together the contributions of a number of outstanding researchers in the field of epoxy resins. Not only emphasizes the chemistry and technology of epoxy resins, but also deals with many industrial applications. Of great value for polymer chemists and technicians, and a wide variety of engineers.

CONTENTS: Introduction to epoxy resins, *C. May.* Synthesis and characteristics of epoxides, *Y. Tanaka, A. Okada, and I. Tomizuka.* Epoxide-curing reactions, *Y. Tanaka and T. Mika.* Curing agents and modifiers, *T. Mika.* Properties of cured resins, *D. Kaelble.* Epoxy-resin adhesives, *A. Lewis and R. Saxon.* Epoxy-resin coatings, *G. Somerville and I. Smith.* Electrical and electronic applications, *A. Breslau.* Epoxy laminates, *J. DelMonte.* Polymer stabilizers and plasticizers, *W. Port.* Analysis of epoxides and epoxy resins, *H. Jahn and P. Goetzky.* Toxicity, hazards, and safe handling, *H. Borgstedt and C. Hine.*

MILLICH and CARREHER *Interfacial Synthesis*

edited by FRANK MILLICH, *Department of Chemistry, University of Missouri, Kansas City,* and CHARLES E. CARREHER, JR., *Department of Chemistry, University of South Dakota, Vermillion*

in preparation. 1973

Summarizes the accomplishments in interfacial synthesis to date. Speculates on mechanism, discusses the complex matter of synthetic control, and points out the beneficial aspects of interfacial synthesis in comparison to alternative methods of synthesis. Of fundamental concern to graduate students and teachers of organic chemistry, physical chemists, macromolecular biochemists, and polymer chemists who engage in interfacial synthesis.

CONTENTS: Stirring in organic chemical synthesis, *J. Rushton.* High-speed stirring in interfacial synthesis, *J. Rushton.* Problems and solutions in kinetics and mechanisms, *J. Bradbury and P. Crawford.* Interface effects on chemical reaction rate, *F. MacRitchie.* Liquid-vapor interfacial polycondensations, *L. Sokolov.* Copolycondensation and macroscopic kinetics, *L. Sokolov and V. Nikonov.* The role of the particle-water interface in polymerization, *J. Gardon.* Biochemical reactions at an interface, *R. Baier and D. Cadenhead.* Commercial application of interfacial synthesis, *E. Oliver and Y. Yen.* Polycarboxylic esters, *S. Temin.* Polycarbonates, *H. Vernaleken.* Polycondensations with carbon suboxide, *I. Daniewska.* Polyamides, *V. Nikonov and V. Savinov.* Polyesteramides, *I. Panayotov.* Polyurethanes, *T. Tanaka and T. Yokoyama.* Polyureas, *K. Stueben and A. Barnabeo.* Polyphosphonates, polyphosphates, and polyphosphites, *F. Millich, J. Teague, L. Lambing, and D. Hackathorn.* Other phosphorus containing polymers, *C. Carreher, Jr.* Organometallic polymers, *C. Carreher, Jr.* Modification of natural polymers by interfacial methods, *M. Horio.* Interfacial modifications of poly(vinyl alcohol) and related polymers, *M. Tsuda.* High temperature resistant polymers made by interfacial polymerization, *H. Mark and S. Atlas.*

MYERS and LONG Characterization of Coatings: Physical Techniques

In 2 Parts

(Treatise on Coatings Series, Volume 2)

edited by RAYMOND R. MYERS, *Paint Research Institute, Kent State University, Ohio,* and J. S. LONG, *Department of Chemistry, University of Southern Mississippi, Hattiesburg*

Part I 696 pages, illustrated. 1969

Part II in preparation. 1973

Explores the scientific frontier that has developed since the appearance of Mattiello's treatise on coatings. Emphasizes the urgent need of the paint industry to master new technological concepts and instrumental techniques to match the rapid pace of development of its products. Written for the working paint scientist, the laboratory assistant, technician, and superintendent. Also a valuable reference for the formulator and personnel engaged in the production of raw materials for the paint industry.

CONTENTS:

Part I: The intrinsic properties of polymers, *A. Tawn.* Surface areas, *D. Gans.* Adhesion of coatings, *A. Lewis and L. Forrestal.* Mechanical properties of coatings, *P. Pierce.* The ultimate tensile properties of paint films, *R. Evans.* Gas chromatography, *J. Haken.* Thermoanalytical techniques, *P. Garn.* Microscopy in coatings and coating ingredients, *W. Lind.* Radioactive isotopes, *G. Coe.* Infrared spectroscopy *C. Smith.* Ultraviolet and visible spectroscopy, *F. Spagnolo and E. Scheffer.* Color of polymers and pigmented systems, *G. Ingle.* Photoelastic coatings, *A. Blumstein.*

Part II: Dielectric properties, *S. Negami.* Gel permeation chromatography, *K. Boni.* Infrared Fourier transform spectroscopy, *M. Low.* Interfacial energetics, *D. Gans.* Nuclear magnetic resonance, *M. Levy.* Particle sizing, *B. DeWitt.* Scanning electron microscopy, *L. Princen.* Solubility, *J. Gordon and J. Teas.* Transport properties, *G. Park.* Viscometry, *K. Oesterle.* X-ray analysis, diffraction, and emission, *R. Scott.*

MYERS and LONG Film-Forming Compositions

In 3 Parts

(Treatise on Coatings Series, Volume 1)

edited by RAYMOND R. MYERS, *Paint Research Institute, Kent State University, Ohio,* and J. S. LONG, *Department of Chemistry, University of Southern Mississippi, Hattiesburg*

Part I 584 pages, illustrated. 1967

Part II 448 pages, illustrated. 1968

Part III 608 pages, illustrated. 1972

Devoted to materials which form, or aid the formation of, continuous films. Discusses vehicles and resins, placing considerable emphasis on procedures for developing a suitable vehicle for conveying dissolved or suspended solids to a substrate and imparting to the surface those protective and decorative properties for which the coating was designed. Of inestimable value to chemists, formulators, laboratory assistants, technicians, and production superintendents of the coatings industry. Also valuable for laboratory personnel of raw material suppliers, chemists in the plastics and similar industries, and as an excellent reference treatise for libraries.

Part I: Acrylic ester emulsions and water-soluble resins, *G. Allyn.* Acrylic ester resins, *G. Allyn.* Alkyd resins, *W. M. Kraft, E. G. Janusz, and D. Sughrue.* Asphalt and asphalt coatings, *S. H. Alexander.* Chlorinated rubber, *H. E. Parker.* Driers, *W. J. Stewart.* Epoxy resin coatings, *G. R. Somerville.* Hydrocarbon resins and polymers, *D. F. Koenecke.* Hydrocarbon solvents, *W. W. Reynolds.* Natural resins, *C. L. Mantell.* Polyethers and polyesters, *A. C. Filson.* Urethane coatings, *A. Damusis and K. C. Frisch.* Vehicle manufacturing equipment, *A. F. Steioff.*

Part II: Styrene-butadiene latexes in protective and decorative coatings, *F. A. Miller.* Starch polymers and their use in paper coating, *T. F. Protzman and R. M. Powers.* Cellulose esters and ethers, *J. B. G. Lewin.* Drying oils — modifications and use, *A. E. Rheineck and R. O. Austin.* Paint and painting in art, *S. Rees Jones.* Rosin and modified rosins and resins, *C. L. Mantell.* Urea and melamine resins, *H. P. Wohnsiedler and W. L. Hensley.* Vinyl resins *W. H. McKnight and G. S. Peacock.* Vinyl emulsions, *H. D. Cogan and A. L. Mantz.*

Part III: Dimer acids in surface coatings, *J. Boylan.* Emulsion technology, *L. Princen.* Phenolic resins for coatings, *S. Richardson and W. Wertz.* Plasticization of coatings, *F. Ball.* Fatty polyamides and their applications in protective coatings, *D. Wheller and D. Peerman.* Polycarbonate resins, *D. Fox and K. Goldblum.* Reactive polyesters, *F. Ball.* Varnishes, *L. Montague.* Reactive silanes as adhesion promoters to hydrophilic surfaces, *E. Plueddemann.* Surface-active agents, *T. Ginsberg.* Shellac, *J. Martin.* Tall oil in surface coatings, *R. Perez.* Silicones in protective coatings, *L. Brown.*

MYERS and LONG   *Formulations*

(Treatise on Coatings Series, Volume 4) edited by RAYMOND R. MYERS, *Department of Chemistry, Kent State University, Ohio,* and J. S. LONG, *University of Southern Mississippi, Hattiesburg*

Part I in preparation. 1973

NEUSE and ROSENBERG
Metallocene Polymers

(Reviews in Macromolecular Chemistry Series, Volume 5, Part I)

by EBERHARD NEUSE, *F. J. Weck Company, City of Industry, California,* and HAROLD ROSENBERG, *Air Force Materials Laboratory, Wright-Patterson Air Force Base, Ohio*

170 pages, illustrated. 1970

Presents a comprehensive and critical account of the progress made in the synthesis and characterization of metallocene polymers.

CONTENTS: Introduction • Macromolecular compounds with pendent metallocenyl groups • Macromolecular compounds with intrachain metallocenylene groups • Conclusions.

O'CONNOR   *Instrumental Analysis of Cotton Cellulose and Modified Cotton Cellulose*

(Fiber Science Series, Volume 3)

edited by ROBERT T. O'CONNOR, *Agricultural Research Service, U.S.D.A., New Orleans, Louisiana*

512 pages, illustrated. 1972

Describes the applications of instrumental procedures specifically developed by the textile chemist to meet today's demands. Particularly geared to textile chemists and others in the textile industry, and to paper and wood manufacturers. Also of interest to polymer chemists, analytical chemists, and other researchers involved with the processes by which fibers are blended and modified.

CONTENTS: Elemental analysis: Detection, identification, and quantitative determination of metals and nonmetallic elements, *R. T. O'Connor.* Infrared spectroscopy and physical properties of cellulose, *C. Y. Liang.* Light microscopy in the study of cellulose, *M. L. Rollins and I. V. de Gruy.* Electron microscopy of cellulose and cellulose derivatives, *M. L. Rollins, A. M. Cannizzaro, and W. R. Goynes.* Instrumental methods in the study of oxidation, degradation, and pyrolysis of cellulose, *P. K. Chatterjee and R. F. Schwenker, Jr.* X-ray diffraction, *V. W. Tripp and C. M. Conrad.* Wide-line nuclear magnetic resonance spectroscopy, *R. A. Pittman and V. W. Tripp.* The infrared spectra of chemically modified cotton cellulose, *R. T. O'Connor.*

PEARL   *The Chemistry of Lignin*

by IRWIN A. PEARL, *The Institute of Paper Chemistry, Appleton, Wisconsin*

360 pages, illustrated. 1967

CONTENTS: Nebulous concept of lignin • Isolation of lignin • Chemical structure of lignin • Biosynthesis and formation of lignin • Reactions of lignin in major pulping and bleaching processes • Chemical reactions of lignin • Physical properties of lignin and its preparations • Biological decomposition of lignin • Thermal decomposition of lignin • Linkage of lignin in the plant • Utilization of lignin and its preparations.

PETERLIN   *Plastic Deformation of Polymers*

edited by A. PETERLIN, *Research Triangle Institute, Research Triangle Park, North Carolina*

318 pages, illustrated. 1971

Explores the actual mechanism of deformation that occurs during the formation of fibers and films. Focuses primarily on three topics: what happens on a molecular, crystalline, and supercrystalline level; how the original structure influences the deformation; and what determines the useful mechanical properties of fibers and films. A valuable tool for the staff of research and development laboratories working with plastics, rubbers, fibers, and films, and for students and faculty involved in material science and polymer chemistry.

CONTENTS: Infrared studies of the role of monoclinic structure in the deformation of polyethylene, *Y. Kikuchi and S. Krimm.* Infrared studies of drawn polyethylene. Part I. Changes in orientation and conformation of highly drawn linear polyethylene, *W. Glenz and A. Peterlin.* Structure of oriented polyacrylonitrile films, *J. L. Koenig, L. E. Wolfram, and J. G. Grasselli.* Morphology and deformation behavior of "row-nucleated" polyoxymethylene film, *C. A. Garber and E. S. Clark.* Plastic deformation of polypropylene. VI. Mechanism and properties, *F. J. Baltá-Calleja and A. Peterlin.* Polyethylene crystallized under the orientation and pressure of a pressure capillary viscometer. Part I., *J. H. Southern and R. S. Porter.* Heat relaxation of drawn polyoxymethylene, *A. Siegmann and P. H. Geil.*

Retraction of cold-drawn polyethylene and polypropylene, *D. Hansen, W. F. Kracke, and J. R. Falender.* Electron spin resonance studies of free radicals in mechanically loaded nylon 66, *G. S. P. Verma and A. Peterlin.* Transition from linear to nonlinear viscoelastic behavior. Part I. Creep of polycarbonate, *I. V. Yannas and A. C. Lunn.* Yielding behavior of glassy polymers. III. Relative influences of free volume and kinetic energy, *K. C. Rusch and R. H. Beck, Jr.* Yielding of quenched and annealed polymethyl methacrylate, *D. H. Ender.* Yield phenomenon in oriented polyethylene terephthalate, *M. Parrish and N. Brown.* Electron paramagnetic resonance investigation of molecular bond rupture due to ozone in deformed rubber, *K. I. DeVries, E. R. Simonson, and M. L. Williams.* Factors affecting the depth of draw in a cold-forming operation, *H. L. Li, P. J. Koch, D. C. Prevorsek, and H. J. Oswald.* Quantitative structural characterization of the mechanical properties of isotactic polypropylene, *R. J. Samuels.*

RAVVE   *Organic Chemistry of Macromolecules: An Introductory Textbook*

by A. Ravve, *Continental Can Company, Chicago, Illinois*

512 pages, illustrated. 1967

CONTENTS: **Part I: Introduction:** Historical introduction and definitions • Physical properties of macromolecules • Molecular weights of polymers • **Part II: Polymerization Reactions-Mechanisms:** Addition polymerization: Mechanism of free-radical polymerization • Ionic polymerization • Polymerization with the aid of complex catalysts • Stereospecific polymerization • Bulk, solution, suspension, and emulsion polymerization. **Part III: Common Addition Polymers:** Macroalkanes • Polymers and copolymers from dienes and polyenes • Styrene and styrene-like polymers and polyacrylics • Halogen-bearing addition polymers and vinyl esters and ethers. **Part IV: Condensation Polymers:** Mechanism of polycondensation reactions • polyesters • Polyamides • Polycarbamates, polyureas, and polycarbodiimides • Phenoplasts • Aminoplasts • Ladder and semiladder polymers. **Part V: Naturally Occurring Polymers:** Polysaccharides • Proteins • Polynucleotides. **Part VI: Reactions of Polymers:** Graft and block copolymers • Reactions of polymers • Degradation of polymers.

REICH and STIVALA   *Autoxidation of Hydrocarbons and Polyolefins: Kinetics and Mechanisms*

by Leo Reich, *Picatinny Arsenal, Dover, New Jersey,* and Salvatore S. Stivala, *Department of Chemistry, Stevens Institute of Technology, Hoboken, New Jersey*

544 pages, illustrated. 1969

CONTENTS: Introduction • Oxidation of simple hydrocarbons in absence of inhibitors and accelerators • Oxidation of simple hydrocarbons in presence of antioxidants • Oxidation of simple hydrocarbons in presence of metal catalysts • Weak chemiluminescence during hydrocarbon autoxidation • Qualitative aspects of autoxidation of saturated polyolefins • Quantitative aspects of autoxidation of saturated polyolefins • Investigation of polyolefin oxidation by various techniques.

REMBAUM and SHEN   *Biomedical Polymers*

edited by Alan Rembaum, *California Institute of Technology, Pasadena* and Mitchel Shen, *University of California, Berkeley*

304 pages, illustrated. 1971

A collection of papers given at the Symposium on Biomedical Polymers held in Pasadena in 1969. Of interest to scientists in polymer chemistry, polymer physics, biochemistry, bioengineering, materials science, pharmacology, and surgery.

CONTENTS: Problems in blood-tissue reactions to polymeric materials, *B. Zweifach.* Past, present, and future of artificial kidney treatment, *B. Barbour.* The chemistry and properties of the medical-grade silicones, *S. Braley.* Correlation of the surface charge characteristics of polymers with their antithrombogenic characteristics, *S. Srinivasan and P. Sawyer.* Persistent polarization in polymers and blood compatibility, *P. Murphy, A. Lacroix, S. Merchant, and W. Bernhard.* Selection, characterization, and biodegradation of surgical epoxies, *A. Cupples and R. Schubert.* Foreign body reactions to plastic implants, *D. Ocumpaugh and H. Lee.* Rapid *in vitro* screening of polymers for bio-compatibility, *C. Homsy, K. Ansevin, W. O'Bannon, S. Thompson, R. Hodge, and M. Estrella.* Improved membranes for hemodialysis, *F. Martin, H. Shuey, and C. Saltonstall, Jr.* Control of polymer morphology for biomedical applications, 1. Hydrophilic polycarbonate membranes for dialysis, *R. Kesting.* Surgical adhesives in ophthalmology, *M. Refojo.* Medical uses for polyelectrolyte complexes, *M. Vogel, R. Cross, H. Bixler, and R. Guzman.* Potentialities of a new class of anticlotting and antihemorrhagic polymers, *T. Yen, M. Daver, and A. Rembaum.* Synthesis and properties of a new class of potential biomedical polymers, *A. Rembaum, S. Yen, R. Landel, and M. Shen.* Recognition polymers, *D. Bradley.* The challenge for high polymers in medicine, surgery, and artificial internal organs, *H. Lee and K. Neville.*

RICHARDSON   *Optical Microscopy for the Materials Sciences*

(Monographs and Textbooks in Material Science Series, Volume 3)

by James H. Richardson, *Aerospace Corporation, El Segundo, California*

(continued)

RICHARDSON *(continued)*

704 pages, illustrated. 1971

Provides in one volume a comprehensive survey of the techniques for preparation and optical examination of specimens in the broad area of materials sciences. A highly useful text for the university or vocational school student studying the microstructure of materials or metallography and also a valuable practical reference for all researchers using microscopy, as well as for engineers and industrial metallographers.

CONTENTS: The Brightfield optical microscope • The microscopy of phase structures • Photomicrography • Photomacrography • Specimen preparation • Reagents and techniques for specimen preparation • Laboratory safety • Examination of the specimen • Accessories • Laboratory design.

ROGERS *Permselective Membranes*

edited by CHARLES E. ROGERS, *Case Western Reserve University, Cleveland, Ohio*

224 pages, illustrated. 1971

Encompasses a broad range of topics pertaining to the expanding field of permselective membranes, including theoretical aspects of transport behavior, new and unusual methods for the preparation or modification of membrane materials, and the effects of experimental conditions on the permselectivity of membranes to both ionic and nonionic penetrants. Of value to biophysicists, polymer chemists, physicists, biochemists, and chemical engineers, as well as biologists and physiologists.

CONTENTS: Transport of dissolved oxygen through silicone rubber membrane, *S. Hwang, T. Tang, and K. Kammermeyer.* Gas transport in segmented block copolymers, *K. Ziegel.* Transport of noble gases in poly(methyl acrylate), *W. Burgess, H. Hopfenberg, and V. Stannett.* Permeation of gases at high pressures, *S. Stern, S. Fang, and R. Jobbins.* Permeation of gases through modified polymer films III. Gas permeability and separation characteristics of gamma–irradiated Teflon FEP copolymer films, *R. Huang and P. Kanitz.* Theoretical interpretation of the effect of mixture composition on separation of liquids in polymers, *M. Fels and R. Huang.* Permselectivity of solutes in homogeneous water-swollen polymer membranes, *H. Yasuda and C. Lamaze.* Ion–exchange selectivity coefficients in the exchange of calcium, strontium, cobalt, nickel, zinc, and cadmium ions with hydrogen ion in variously cross–linked polystyrene sulfonate cation exchangers at $25°C$, *M. Reddy and J. Marinsky.* Ion and water transport through permselective

membranes, *N. Lakshminarayanaiah.* Permeability of cellulose acetate membranes to selected solutes, *H. Lonsdale, B. Cross, F. Graber, and C. Milstead.* Transport through permselective membranes, *C. Rogers and S. Sternberg.*

SCHEY *Metal Deformation Processes: Friction and Lubrication*

(Monographs and Textbooks in Material Science Series, Volume 1)

edited by JOHN A. SCHEY, *University of Illinois at Chicago Circle*

822 pages, illustrated. 1970

A comprehensive treatment of all aspects of friction and lubrication in metal deformation processes. Of aid to metallurgists, mechanical engineers, chemists, and physicists.

CONTENTS: Background and system of approach, *J. Schey.* Friction effects in metalworking processes, *J. Schey.* Friction, lubrication, and wear mechanisms, *C. Riesz.* Lubricants, *C. Riesz.* Lubricant properties and their measurements, *J. Schey.* Rolling lubrication, *J. Schey.* Wire drawing lubrication, *J. Newnham.* Hot extrusion lubrication, *S. Kalpakjian.* Forging lubrication, *S. Kalpakjian.* Cold forging and cold extrusion lubrication, *J. Newnham.* Sheet metal working lubrication, *J. Newnham.*

SEGAL *High-Temperature Polymers*

edited by CHARLES L. SEGAL, *North American Aviation, Inc., Canoga Park, California*

208 pages, illustrated. 1967

CONTENTS: Introduction, *C. L. Segal.* Thermally stable polymers with aromatic recurring units, *C. S. Marvel.* Inorganic polymer chemistry, *J. R. Van Wazer.* Kinetics and gaseous products of thermal decomposition of polymers, *H. L. Friedman.* Studies of stability of condensation polymers in oxygen-containing atmospheres, *R. T. Conley.* Thermal degradation of polymers. III: Mass spectrometric thermal analysis of condensation polymers, *G. P. Shulman.* Viscoelastic relaxation mechanism of inorganic polymers. V: Counterion effects in bulk polyelectrolytes, *A. Eisenberg, S. Saito, and T. Sasada.* Thermal stability of carborane-containing polymers, *J. Green and N. Mayes.* Synthesis and thermal stability of structurally related aromatic Schiff bases and acid amides, *A. D. Delman, A. A. Stein, and B. B. Simms.* New high-temperature polymers. II: Ordered aromatic copolyamides containing fused and multiple ring systems, *F. Dobinson and J. Preston.* Synthesis of fusible branched polyphenylenes, *N. Bilow and L. J. Miller.*

SEGAL, SHEN, and KELLEY
Polymers in Space Research

edited by CHARLES L. SEGAL, *Whittaker Corporation, San Diego*, MITCHEL SHEN, *University of California, Berkeley*, and FRANK N. KELLEY, *Air Force Rocket Propulsion Laboratory, Edwards, California.* Associate Editors: GEORGE F. PEZDIRTZ, *NASA Langley Research Center, Hampton, Virginia*, and W. DAVID ENGLISH, *Astropower Laboratory, McDonnell Douglas Aeronautics Company, Huntington Beach, California*

480 pages, illustrated. 1970

CONTENTS:

Part I: Recent Developments in the Synthesis, Characterization, and Evaluation of Thermally Stable Polymers

Introduction, *C. Segal and G. Pezdirtz.* Aromatic polymers: Single- and double-stranded chains, *J. Stille.* Thermally stable spiropolymers, *J. Hodgkin and J. Heller.* Isomeric and substituent effects in some dibenzoylbenzene-diamine polymers, *A. Volpe, L. Kaufman, and R. Dondero.* Arylsulfimide polymers. III. The syntheses of some monomeric aryl-1,2-disulfonic acids and derivatives, *G. D'Alelio, Y. Giza, and D. Feigl.* Properties of heterocyclic condensation polymers, *G. Berry and T. Fox.* Relative thermophysical properties of some polyimidazopyrrolones, *R. Jewell.* Thermal decomposition of polyimides in vacuum, *T. Johnston and C. Gaulin.* Thermomechanical behavior of an aromatic polysulfone, *J. Gillham, G. Pezdirtz, and L. Epps.* Panel discussion on thermally stable polymers, *C. Segal, J. Stille, G. Pezdirtz, G. D'Alelio, H. Levine, and W. Gibbs.*

Part II: Properties of Polymers at Low Temperatures

Introduction, *M. Shen and W. English.* Relaxation behavior of polymers at low temperatures, *J. Sauer and R. Saba.* Thermal properties of polymers at low temperatures, *W. Reese.* Multiple transitions in polyvinyl alkyl ethers at low temperatures, *W. Schell, R. Simha, and J. Aklonis.* Internal friction study of diluent effect on polymers at cryogenic temperatures, *M. Shen, J. Strong, and H. Schlein.* Stress-strain behavior of adhesives in a lap joint configuration at ambient and cryogenic temperatures, *G. Tiezzi and H. Doyle.* Some properties of nitroso rubbers in fluorine at ambient and cryogenic temperatures, *S. Toy, W. English, W. Crane, and M. Toy.* Cryogenic properties of a polyurethane adhesive, *R. Robbins.* Some effects of structure on a polymer's performance as a cryogenic adhesive, *R. Gosnall and H. Levine.*

Part III: Solid Propellants

Introduction, *F. Kelley.* Recent developments in solid-propellant binders, *H. Marsh, Jr.* Saturated hydrocarbon polymers for solid rocket propellants, *A. Di Milo and D. Johnson.* Preparation and curing of poly (perfluoroalkylene oxides), *J. Zollinger, J. Throckmorton, S. Ting, R. Mitsch, and D. Elrick.* Functionality and functionality distribution measurements of binder prepolymers, *A. Muenker and B. Hudson, Jr.*

SERAFINI and KOENIG
Cryogenic Properties of Polymers

edited by TITO T. SERAFINI, *NASA-Lewis Research Center, Cleveland*, and JACK L. KOENIG, *Case Western Reserve University, Cleveland, Ohio*

312 pages, illustrated. 1968

CONTENTS: Cryogenic positive expulsion bladders, *R. F. Lark.* Adhesives for cryogenic applications, *L. M. Roseland.* Glass-, boron-, and graphite-filament-wound resin composites and liners for cryogenic pressure vessels, *M. P. Hanson.* Mechanical behavior of poly (ethylene terephthalate), *I. M. Ward.* Effect of film processing on cryogenic properties of poly (ethylene terephthalate), *R. E. Eckert and T. T. Serafini.* Mechanical properties of epoxy resins and glass/epoxy composites at cryogenic temperatures, *L. M. Soffer and R. Molho.* Mechanical relaxation of poly-4-methyl-pentene-1 at cryogenic temperatures, *M. Takayanagi and N. Kawasaki.* Transitions in glasses at low temperatures, *R. A. Haldon, W. J. Schell, and R. Simha.* Mechanical behavior of poly (ethylene terephthalate) at cryogenic temperatures, *C. D. Armeniades, I. Kuriyama, J. M. Roe, and E. Baer.* Infrared studies of chain folding in polymers II. Poly (ethylene terephthalate), *J. L. Koenig and M. J. Hannon.* Crystallization of poly (ethylene terephthalate) from the glassy amorphous state, *G. S. Y. Yeh and P. H. Geil.* Strain-induced crystallization of poly (ethylene terephthalate), *G. S. Y. Yeh and P. H. Geil.* Molecular motion in polytetrafluoroethylene at cryogenic temperatures, *E. S. Clark.* Synthesis of ultrahigh molecular weight poly (ethylene terephthalate), *L.-C. Hsu.* Development of vulcanizable elastomers suitable for use in contact with liquid oxygen, *P. D. Schuman, E. C. Stump, and G. Westmoreland.* Synthesis of fluorinated polyurethanes, *R. Gosnell and J. Hollander.*

SKEIST   Reviews in Polymer Technology

edited by IRVING SKEIST, *Skeist Laboratories, Inc., Livingston, New Jersey*

260 pages, illustrated. 1972

Consists of intensive, up-to-date reviews on various aspects of polymer technology. Directed to chemists, engineers, technicians, commercial planners, and others working in the polymers and plastics fields.

CONTENTS: Coupling agents as adhesion promoters, *P. Cassidy and W. Yager.* Processing powdered polyethylene, *A. Zimmerman.* Plastics and other polymers in building, *I. Skeist and J. Miron.* Fire retardance of polymeric ma-

(continued)

SKEIST *(continued)*

terials, *I. Einhorn*. Recent advances in photo-cross–linkable polymers, *G. Delzenne*. Organic colorants for polymers, *T. Reeve*.

SOLOMON   Step-Growth Polymerizations

(Kinetics and Mechanisms of Polymerization Series, Volume 3)

edited by DAVID H. SOLOMON, *C.S.I.R.O., Melbourne, Australia*

416 pages, illustrated. 1972

Presents a critical and constructive assessment of developments in step-growth polymerization and considers the application of theoretical concepts to commercial systems. Highly recommended to students of polymer science and researchers working in the area of step-growth polymerization, including polymer chemists and other scientists in the paint, coatings, and plastics industries.

CONTENTS: Polyesterification, *D. H. Solomon*. Polyamides, *D. C. Jones and T. R. White*. Polyurethanes: The chemistry of the diisocyanate-diol reaction, *D. J. Lyman*. Cyclopolycondensation, *P. M. Hergenrother*. The reactions of formaldehyde with phenols, melamine, aniline, and urea, *M. F. Drum and J. R. LeBlanc*. Diels-Alder polymerization, *W. J. Bailey*. Inorganic polymers, *J. R. MacCallum*.

STEWART   Infrared Spectroscopy:
Experimental Methods and Techniques

by JAMES E. STEWART, *Durrum Instrument Corporation, Palo Alto, California*

656 pages, illustrated. 1970

A guide to instrumentation and experimental methods and techniques for the infrared spectroscopist. Primarily for those involved with spectroscopy research and for graduate students in chemistry and physics interested in spectroscopy.

CONTENTS: Infrared spectroscopy • The infrared spectrophotometer • Elements of geometric optics • Elements of physical optics • Optical components of infrared spectrophotometers • Optical systems of infrared spectrophotometers • Slit functions and spectral modulation transfer functions of monochromators • Interference spectroscopy • Mechanics of infrared spectrophotometers • Elements of electronics • Infrared detectors • Electronic systems of infrared spectrophotometers • Electromechanical transfer functions of infrared spectrophotometers • Photometric accuracy of infrared spectrophotometers • Experimental methods of infrared transmission spectroscopy • Experimental methods of infrared reflection spectroscopy • Experimental methods of infrared emission spectroscopy • Advanced methods of infrared spectroscopy.

SZEKELY   Blast Furnace Technology:
Science and Practice

edited by JULIAN SZEKELY, *State University of New York, Buffalo*

414 pages, illustrated. 1972

Represents the efforts of academic and industrial researchers, metallurgists, plant operators, and designers concerned with the most up-to-date aspects of ironmaking technology. Valuable reading for all production engineers, plant operators, and designers concerned with ironmaking technology, and also of importance to metallurgical engineers, research scientists—chemists, physicists, engineers—and students in this field.

CONTENTS: Single particle studies applied to direct reduction and blast furnace operations, *R. Bleifuss*. Structural effects in gas–solid reactions, *J. Szekely and J. Evans*. The use of catalysts to enhance the rate of Boudouard's reaction in direct reduction metallurgical processes, *Y. Rao and B. Jalan*. Thirty psi high top–gas pressure operation at NSC Nagoya works, *T. Yatsuzuka, Y. Yamada, and A. Tayama*. Practical application of mathematical models in ironmaking, *D. Christie, C. Kearton, and R. Thomas*. A mass–transport model of erosion of the carbon hearth of the iron blast furnace, *J. Elliott and J. Popper*. Contribution to the study of the reaction mechanism occurring in high temperature zone of the blast furnace, *R. Vidal and A. Poos*. The place of direct reduction in a modern blast furnace—BOF plant, *J. Peart and D. George*. The blast furnace control problem, *J. A. Laslo*. Modern blast furnace design in Germany, *F. Lenger*. The blast furnace—a transition, *F. Berczynski*. Projected performance of a blast furnace with prereduced burdens, *J. Agarwal*.

SZEKELY   The Steel Industry and the Environment

edited by JULIAN SZEKELY, *Center for Process Metallurgy, State University of New York, Buffalo*

312 pages, illustrated. 1973

Contains the proceedings of the Second C. C. Furnas Memorial Conference on The Steel Industry and The Environment held at the State University of New York at Buffalo in November, 1971. Brings together authors representing different viewpoints on the questions raised in examining the interaction of the steel industry and the environment. Of utmost importance to plant operators, designers, metallurgists, environmental scientists, and all others concerned with the impact of the steel industry on the environment.

CONTENTS: The role of the federal government in environmental pollution control, *K. Johnson.* Control of air pollution in the British iron and steel industry, *F. Ireland.* Health and the steel industry environment, *K. Spring.* The economic impact of the installation and operation of pollution abatement devices, *J. Barker.* Desulfurization of coke oven gas: Technology, economics, and regulatory activity, *R. Dunlap, W. Gorr, and M. Massey.* Treatment of cold-mill wastewaters by ultrahigh-rate filtration, *C. Symons.* Experience with pollution abatement, *C. Black and W. Sebesta.* The interaction of the socioeconomic and ecological environment in American steel mill towns, *L. Thaxton and R. Genton.* Plant availability versus clean air: An economic dilemma that can be solved, *R. Heller.* A survey of wastewater treatment techniques for steel mill effluents, *T. Centi.* Emission of sulfurous gases from blast-furnace slags, *R. Kaplan and G. Ringstorff.* Treatment of waste gases from the basic oxygen furnace in West Germany, *E. Weber.* On the oxidation of cyanides in the stack region of the blast furnace, *H. Sohn and J. Szekely.* Reclaimed scrap and solid metallics for steelmaking, *J. Elliott.*

TALLAN Electrical Conductivity in Ceramics and Glass

(Ceramics and Glass: Science and Technology Series, Volume 4)

edited by NORMAN M. TALLAN, *Aerospace Research Laboratories, Wright-Patterson Air Force Base, Ohio*

in preparation. 1973

A text which thoroughly describes several aspects of the electrical conductivity of ceramic materials, and discusses their conductivity, physical dependence on their electronic and ionic defect structures, and the transport mechanisms by which charge and mass move through ceramic materials. Additionally stressed is the use of conductivity measurements to characterize the defect structure and transport properties of ceramics. A great aid to advanced students in materials science, ceramics, and glass.

CONTENTS: General concepts of electrical transport, *D. Adler.* Experimental techniques, *R. Blumenthal and M. Seitz.* Defect structure of ceramic materials, *R. Brook.* Electronic conduction mechanisms, *I. Bransky and J. Wimmer.* Controlled valency effects in electronic conductors, *J. Wagner.* Highly conducting ceramics and the conductor-insulator transition, *J. Honig and R. Vest.* Ionic conductivity and electrochemistry of crystalline ceramics, *J. Patterson.* Conductivity of glass and other amorphous materials, *J. Mackenzie.* Microstructural and polyphase effects, *J. Wimmer and H. Graham.*

TSURUTA and O'DRISCOLL Structure and Mechanism in Vinyl Polymerization

edited by TEIJI TSURUTA, *Department of Synthetic Chemistry, Faculty of Engineering, University of Tokyo,* and KENNETH F. O'DRISCOLL, *Department of Chemical Engineering, State University of New York, Buffalo*

552 pages, illustrated. 1969

Presents a general survey of studies on this subject in terms of physical organic chemistry. Topics are organized to focus on the most important chemical features of vinyl compounds and their response to variations in chemical and physical circumstances.

CONTENTS: Historical development of the theory of the reactivity of vinyl monomers, *M. Imoto.* Structure and reactivity of vinyl monomers, *T. Tsuruta.* Initiation in free radical polymerization, *K. F. O'Driscoll and P. Ghosh.* Termination mechanism in radical polymerization, *A. M. North and D. Postlethwaite.* Organometallic compounds as radical-type initiators for vinyl polymerization, *S. Inoue.* Heterogeneous metal peroxides, *T. Otsu.* Polymerization of α, β-disubstituted olefins, *Y. Minoura.* Polymerization of α, β-unsaturated carbonyl compounds, *D. M. Wiles.* Cationic polymerization of vinyl monomers by metal alkyl catalysts, *T. Saegusa.* Rate constants of elementary reactions in cationic polymerization, *T. Higashimura.* Elementary steps in anionic vinyl polymerization, *J. Smid.* Molecular rearrangements in polymerization of vinyl monomers, *A. D. Ketley and L. P. Fisher.*

VOGL Polyaldehydes

edited by OTTO VOGL, *Central Research Division, E. I. du Pont de Nemours & Company, Wilmington, Delaware*

152 pages, illustrated. 1967

CONTENTS: Preface, *O. Vogl.* Polyaldehydes: introduction and brief history, *O. Vogl.* Polymerization of formaldehyde, *N. Brown.* Polymerization and copolymerization of trioxane, *M. B. Price and F. B. McAndrew.* Polymerization of aliphatic aldehydes, *O. Vogl.* Polymers of haloaldehydes, *I. Rosen.* NMR studies of polyaldehydes, *E. G. Brame, Jr. and O. Vogl.* Polymerization of fluorothiocarbonyl compounds, *W. H. Sharkey.* Crystal structure of polyaldehydes, *P. Corradini.* Morphology of polyoxymethylene, *P. H. Geil.*

VOGL and FURUKAWA Polymerization of Heterocyclics

edited by OTTO VOGL, *Department of Polymer Science and Engineering, Univer-*

(continued)

VOGL and FURUKAWA *(continued)*

sity of Massachusetts, Amherst, and JUNJI FURUKAWA, *Kyoto University, Japan*

216 pages, illustrated. 1973

Reviews the polymerization of cyclic ethers and thio ethers, lactones, and lactams. Also covers preparation, polymerization, and properties of perfluoro epoxides, kinetics of cyclic ether polymerization, and the influence of ring strain on the rate of polymerization and living polymers based on cationic ring opening polymerization. Valuable to scientists interested in polymer science, heterocyclic chemistry, polymer engineering, and materials science.

CONTENTS: Introduction, *J. Furukawa*. Polymerization of cyclic ethers, *T. Saegusa*. Polymerization of perfluoro epoxides, *H. Eleuterio*. Specific nature of the polymerization of heterocyclics, *N. Enikolpoyan*. New trioxane copolymers, *H. Cherdron*. Alkylene sulfide polymerizations, *F. Lautenschlaeger*. Lactone polymerization and polymer properties, *G. Brode and J. Koleske*. Lactam polymerization, *J. Sebenda*.

WALKER and THROWER *Chemistry and Physics of Carbon: A Series of Advances*

a series edited by PHILIP L. WALKER and PETER A. THROWER, *Department of Material Sciences, Pennsylvania State University, University Park*

Vol. 1 400 pages, illustrated. 1965
Vol. 2 400 pages, illustrated. 1966
Vol. 3 464 pages, illustrated. 1968
Vol. 4 416 pages, illustrated. 1968
Vol. 5 400 pages, illustrated. 1969
Vol. 6 368 pages, illustrated. 1970
Vol. 7 424 pages, illustrated. 1970
Vol. 8 480 pages, illustrated. 1973
Vol. 9 272 pages, illustrated. 1973
Vol. 10 288 pages, illustrated. 1973
Vol. 11 in preparation. 1974

CONTENTS:

Volume 1: Dislocations and stacking faults in graphite, *S. Amelinckx, P. Delavignette, and M. Heerschap*. Gaseous mass transport within graphite, *G. F. Hewitt*. Microscopic studies of graphite oxidation, *J. M. Thomas*. Reactions of carbon with carbon dioxide and steam, *S. Ergun and M. Mentser*. Formation of carbon from gases, *H. B. Palmer and C. F. Cullis*. Oxygen chemisorption effects on graphite thermoelectric power, *P. L. Walker, Jr., L. G. Austin, and J. J. Tietjen*.

Volume 2: Electron microscopy of reactivity changes near lattice defects in graphite, *G. R. Hennig*. Porous structure and adsorption properties of active carbons, *M. M. Dubinin*. Radiation damage in graphite, *W. N. Reynolds*. Adsorption from solution by graphite surfaces, *A. C. Zettlemoyer and K. S. Narayan*. Electronic transport in pyrolytic graphite and boron alloys of pyrolytic graphite, *C. A. Klein*. Activated diffusion of gases in molecular-sieve materials, *P. L. Walker, Jr., L. G. Austin and S. P. Nandi*.

Volume 3: Nonbasal dislocations in graphite, *J. M. Thomas and C. Roscoe*. Optical studies of carbon, *S. Ergun*. Action of oxygen and carbon dioxide above 100 millibars on "pure" carbon, *F. M. Lang and P. Magnier*. X-ray studies of carbon, *S. Ergun*. Carbon transport studies for helium-cooled high-temperature nuclear reactors, *M. R. Everett, D. V. Kinsey, and E. Römberg*.

Volume 4: X-ray diffraction studies on carbon and graphite, *W. Ruland*. Vaporization of carbon, *H. B. Palmer and M. Shelef*. Growth of graphite crystals from solution, *S. B. Austerman*. Internal friction studies on graphite, *T. Tsuzuku and M. H. Saito*. Formation of some graphitizing carbons, *J. D. Brooks and G. H. Taylor*. Catalysis of carbon gasification, *P. L. Walker, Jr., M. Shelef, and R. A. Anderson*.

Volume 5: Deposition, structure and properties of pyrolytic carbon, *J. C. Bokros*. The thermal conductivity of graphite, *B. T. Kelly*. The study of defects in graphite by transmission electron microscopy, *P. A. Thrower*. Intercalation isotherms on natural and pyrolytic graphite, *J. G. Hooley*.

Volume 6: Physical adsorption of gases and vapors of graphitized carbon blacks, *N. N. Avgul and A. V. Kiseleyv*. Graphitization of soft carbons, *J. Maire and J. Méring*. Surface complexes on carbons, *B. R. Puri*. Effects of reactor irradiation on the dynamic mechanical behavior of graphites and carbons, *R. E. Taylor and D. E. Kline*.

Volume 7: The kinetics and mechanism of graphitization, *D. B. Fischbach*. The kinetics of graphitization, *A. Pacault*. Electronic properties of doped carbons, *A. M. Marchand*. Positive and negative magnetoresistances in carbons, *P. Delhaes*. The chemistry of the pyrolytic conversion of organic compounds to carbon, *E. Fitzer, K. Mueller and W. Schaefer*.

Volume 8: The electronic properties of graphite, *I. Spain*. Surface properties of carbon fibers, *D. McKee and V. Mimeault*. The behavior of fission products captured in graphite by nuclear recoil, *S. Yajima*.

Volume 9: Carbon fibers from rayon presursors, *R. Bacon*. Control of structure of carbon for use in bioengineering, *J. Bokros, L. LaGrange, and F. Schoen*. Deposition of pyrolytic carbon in porous solids, *W. Kotlensky*.

Volume 10: The thermal properties of graphite, *B. Kelly and R. Taylor*. Lamellar reactions in graphitizable carbons, *M. Robert, M. Oberlin, and J. Mering*. Methods and mechanisms of growth of synthetic diamond, *F. Bundy, H. Strong, and R. Wentorff, Jr*.

† *Volume edited by Philip L. Walker*

Volume 11: Structure and physical properties of carbon fibers, *W. Reynolds.* Highly oriented pyrolytic graphite, *A. Moore.* Evaporated carbon films, *I. McLintock and J. Orr.* Deformation mechanisms in carbons, *G. Jenkins.*

WARD   Chemical Modification of Papermaking Fibers

(Fiber Science Series, Volume 4)
by KYLE WARD, JR., *Institute of Paper Chemistry, Appleton, Wisconsin*
256 pages, illustrated. 1973

Bridges the gap between research and industrial applications in the field of chemical modification of papermaking fibers. Deals with the chemical changes which produce new or improved properties in paper products. Of particular importance to researchers and technologists in the paper, textile, and related industries, and students of polymer and organic chemistry.

CONTENTS: Introduction • Esterification • Etherification • Oxidation • Crosslinking • Graft polymerization onto cellulose.

WASLEY   Stress Wave Propagation in Solids: An Introduction

(Monographs and Textbooks in Material Science Series, Volume 5)
by RICHARD J. WASLEY, *Department of Chemistry, University of California, Livermore*
344 pages, illustrated. 1973

Provides the fundamentals necessary for the study of the propagation of short duration, high–intensity, nonelastic, mechanical stress disturbances in solids. The first part of the book treats some of the dynamic analyses of elastic solid media which obey Hooke's law. The last section discusses some of the theoretical and experimental aspects of one–dimensional stress waves and shock loading and response. For scientists and engineers desiring further knowledge in the field of stress wave propagation. Also of interest to advanced undergraduate and graduate students of engineering and physics.

CONTENTS: Elasticity: Quasistatic and dynamic response • Wave propagation in extended media • Wave propagation in semi-extended media: Reflection and refraction • Wave propagation in circular cylindrical rods • Selected applications of concepts of elasticity • Nonelastic material behavior • One–dimensional stress wave investigations • Nonelastic (shock) one–dimensional strain wave investigations.

WEINBERG   Tools and Techniques in Physical Metallurgy

In 2 Volumes

edited by FRED WEINBERG, *University of British Columbia, Vancouver*

Vol. 1   416 pages, illustrated. 1970
Vol. 2   376 pages, illustrated. 1970

Aids the non-specialist in understanding and making use of the new instruments and techniques of physical metallurgy.

CONTENTS:

Volume 1: Temperature measurement, *R. Bedford, T. Dauphinee, and H. Preston-Thomas.* X-ray diffraction, *C. M. Mitchell.* Crystal growth and alloy preparation, *F. Weinberg and J. T. Jubb.* Quantitative metallography, *J. R. Blank and T. Gladman.* Metallography, *H. E. Knechtel, W. F. Kindle, J. L. McCall, and R. D. Buchheit.*

Volume 2: Electron microscopy, *E. Smith.* Scanning electron microscopy, *O. Schaaber.* Field-ion microscopy, *B. Ralph.* Thermionic-emission microscopy, *W. L. Grube and S. R. Rouze.* Electron-probe microanalysis, *L. C. Brown and H. Thresh.* Emission spectrography and atomic absorption spectrophotometry, *G. L. Mason.*

WILSON   Radiation Chemistry of Monomers-Polymers-Plastics

by JOSEPH E. WILSON, *Department of Chemistry, Bishop College, Dallas, Texas*
in preparation. 1974

Provides an up-to-date survey of the radiation chemistry of monomers, polymers, and plastics. Gives essential information on radiation properties, measurement, and detection, and the primary chemical results of the interaction of radiation with matter. Of particular interest to polymer and radiation chemists.

CONTENTS: (tentative): Types and sources of radiation • Fundamental effects of the irradiation of matter • Short-term chemical effects of radiation absorption • Radiation chemistry of small molecules • Radiolytic polymerization in homogeneous systems • Radiolytic polymerization in the solid state • Radiation-induced polymerization in thermosetting, polyester, and emulsion systems • Irradiation of polymers: Crosslinking versus scission • Radiolytic grafting of monomers on polymeric films • Radiolytic grafting on fibers.

YOCUM and NYQUIST   Functional Monomers: Their Preparation, Polymerization, and Applications

In 2 Volumes

edited by RONALD H. YOCUM, *The Dow Chemical Company, Freeport, Texas,* and

(continued)

YOCUM and NYQUIST *(continued)*

EDWIN B. NYQUIST, *The Dow Chemical Company, Midland, Michigan*

Vol. 1 712 pages, illustrated. 1973
Vol. 2 321 pages, illustrated. 1973

A practical reference work which deals with functional monomers. Presents a broad technical background on the preparation and polymerization of individual functional monomers and their applications to various areas of industry. Of special interest to both academic and industrial chemists, particularly those working in the paint, coatings, and textile industry.

CONTENTS:
Volume 1: Acrylamide and other alpha, beta, and unsaturated acids, *D. C. MacWilliams.* Reactive halogenated monomers, *C. F. Raley and R. J. Dólinski.* Hydroxy monomers, *E. B. Nyquist.* Sulfonic acids and sulfonate monomers, *D. A. Kangas.*

Volume 2: Reactive heterocyclic monomers, *D. Tomalia.* Acidic monomers, *L. Luskin.* Basic monomers, vinylpyridines and aminoalkyl (METH) acrylates, *L. Luskin.*

ZIEF *Purification of Inorganic and Organic Materials: Techniques of Fractional Solidification*

edited by MORRIS ZIEF, *J. T. Baker Chemical Co., Phillipsburg, New Jersey*

340 pages, illustrated. 1969

Of interest to the chemist, chemical engineer, and metallurgist.

CONTENTS: Analysis of ultrapure materials, *C. L. Grant.* Optical-emission spectrochemical analysis—arc, spark, and flame, *C. L. Grant.* Spark-source mass spectrography, *P. R. Kennicott.* Atomic-absorption spectroscopy, *J. W. Robinson.* Infrared spectrophotometry, *K. E. Stine and W. F. Ulrich.* Gas-liquid chromatography, *R. A. Keller.* Differential thermal analysis and differential scanning calorimetry, *E. M. Barrall, II, and J. F. Johnson.* Electrical resistance-ratio measurement, *G. T. Murray.* Reduction of cyclohexane content of benzene under steady flow conditions, *J. D. Henry, Jr., M. D. Danyi, and J. E. Powers.* Purification of aromatic amines, *B. Pouyet.* The freezing staircase method, *C. P. Saylor.* Purification of aluminum, *J. L. Dewey.* Concentration of humic acids in natural waters, *J. Shapiro.* Fractionation of polystyrene, *J. D. Loconti.* Purification and growth of large anthracene crystals, *J. N. Sherwood.* Purification of indium antimonide, *A. R. Murray.* Purification of alkaline iodides (KI, RbI, CsI), *D. Ecklin.* Zone melting of metal chelate systems, *K. Ueno, H. Kobayashi, and H. Kaneko.* Purification of dienes, *R. Kieffer.* Purification of kilogram quantities of an organic compound, *J. C. Maire and M. Delmas.* Rapid

purification of organic substances, *M. J. van Essen, P. F. J. van der Most, and W. M. Smit.* Investigation of zone-melting purification of gallium trichloride by a radiotracer method, *W. Kern.* Purification of potassium chloride by radio-frequency heating, *R. Warren.* Purification of a metal by electron-beam heating, *R. E. Reed and J. C. Wilson.* Heating by hollow-cathode gas discharge, *W. Class.* Continuous zone refining of benzoic acid, *J. K. Kennedy and G. H. Moates.* Purification of naphthalene in a centrifugal field, *E. L. Anderson.* Zone-melting chromatography of organic mixtures, *H. Plancher, T. E. Cogswell and D. R. Latham.* The concentration of flavors at low temperature, *M. T. Huckle.* Containers for pure substances, *E. C. Kuehner and D. H. Freeman.*

ZIEF and SPEIGHTS *Ultrapurity: Methods and Techniques*

edited by MORRIS ZIEF, *J. T. Baker Chemical Co., Philipsburg, New Jersey,* and ROBERT M. SPEIGHTS, *American Metal Climax, Inc., Golden, Colorado*

720 pages, illustrated. 1972

Brings together for the first time the four essential and interrelated parameters of ultrapurity: preparation, handling, containment, and analysis. Reflects the continuing progress in the preparation of ultrapure chemicals, the explosive growth in developments pertaining to the handling and containment of these materials, as well as the necessity for complete analysis.

Directed to all those working in research, development, or analysis of ultrapure products.

CONTENTS: Purification of alkali halides, *F. Rosenberger.* Purification of organic solvents by frontal-analysis chromatography, *H. Engelhardt.* The preparation of pure sodium and potassium, *R. L. McKisson.* Sublimation of phosphorus pentoxide, *R. D. Mounts.* Purification of proteins by membrane ultrafiltration, *G. J. Fallick.* The purification of p-xylene by partial freezing, *J. R. Gruden and M. Zief.* Purification of isopropylbenzene by preparative gas-liquid chromatography, *J. R. Gruden and M. Zief.* The preparation of ultrapure chemicals by fractional distillation, *H. Plancher and W. E. Haines.* Purification by dry-column chromatography, *F. M. Rabel.* Preparation of ultrapure water, *V. C. Smith.* Preparation and characterization of cholesterol, *I. L. Shapiro.* Contamination problems in trace-element analysis and ultrapurification, *D. E. Robertson.* Airborne contamination, *J. A. Paulhamus.* Glass containers for ultrapure solutions, *P. B. Adams.* Vitreous silica, *G. Hetherington and L. W. Bell.* Ceramics, *C. Garnsworthy.* High-purity chemicals—a challenge to practical analysis, *A. J. Barnard, Jr.* Emission spectroscopy, *E. C. Snooks.* Flame spectrophotometric trace analysis, *D. C. Burrell.* Neutron-activation analysis, *J. J. Kelly.* Visible spectrophotometry, *R. H.*

Weiss. Coulometric titration, *G. W. Higgins*. Information sources for ultrapurification and characterization, *T. E. Connolly*.

ZIEF and WILCOX Fractional Solidification

edited by MORRIS ZIEF, *J. T. Baker Chemical Company, Phillipsburg, New Jersey*, and WILLIAM R. WILCOX, *Aerospace Corporation, Los Angeles*

736 pages, illustrated. 1967

CONTENTS: Introduction, *W. R. Wilcox*. **Part I: Basic Principles:** Phase diagrams, *G. M. Wolten and W. R. Wilcox*. Mass transfer in fractional solidification, *W. R. Wilcox*. Constitutional supercooling and microsegregation, *G. A. Chadwick*. Polyphase solidification, *G. A. Chadwick*. Heat transfer in fractional solidification, *W. R. Wilcox*. **Part II: Laboratory Scale Apparatus:** Laboratory scale apparatus, *E. A. Wynne and M. Zief*. Batch zone melting, *E. A. Wynne*. Progressive freezing, *D. Richman, E. A.*

Wynne, and F. D. Rosi. Continuous-zone melting, *J. K. Kennedy and G. H. Moates*. Column crystallization, *R. Albertins, W. C. Gates, and J. E. Powers*. Zone precipitation and allied techniques, *I. A. Eldib*. **Part III: Industrial Scale Equipment:** Proabd refiner, *J. G. D. Molinari*. Newton Chambers' process, *J. G. D. Molinari*. Rotary-drum technique, *J. C. Chaty*. Phillips fractional-solidification process, *D. L. McKay*. Desalination by freezing, *J. C. Orcutt*. **Part IV: Applications:** Ultrapurification, *J. K. Kennedy, and G. H. Moates*. Ultrapurity in pharamaceuticals, *P. Jannke*. Ultrapurity in electronic materials, *J. K. Kennedy and G. H. Moates*. Ultrapurity in materials research, *J. K. Kennedy, G. H. Moates, and W. R. Wilcox*. Ultrapurity in crystal growth, *G. H. Moates and J. K. Kennedy*. Bulk purification, *J. D. Loconti*. Analytical applications of fractional solidification, *A. S. Yue*. Materials preparation, *D. Richman and F. D. Rosi*. **Part V: Economics:** Economics of fractional solidification, *J. C. Chaty and W. R. Wilcox*. **Part VI: Appendix:** Introduction, *M. Zief and C. E. Shoemaker*. Survey of inorganic materials, *C. E. Shoemaker and R. L. Smith*. Survey of organic materials, *M. Zief*.

OTHER BOOKS OF INTEREST

CUTLER and DAVIS Detergency: Theory and Test Methods

In 2 Parts

(Surfactant Science Series, Volume 5)

edited by W. G. CUTLER, and R. DAVIS, *Whirlpool Corporation, Benton Harbor, Michigan*

Part 1 464 pages, illustrated. 1972
Part 2 in preparation. 1973

JUNGERMANN Cationic Surfactants

(Surfactant Science Series, Volume 4)

edited by ERIC JUNGERMANN, *Armour-Dial, Inc., Chicago*

672 pages, illustrated. 1970

LINFIELD Anionic Surfactants

(Surfactant Science Series)

edited by WARNER M. LINFIELD, *U.S. Department of Agriculture, Philadelphia, Pennsylvania*

in preparation. 1974

MATTSON and MARK
Activated Carbon: Surface Chemistry and Adsorption from Solution

by JAMES S. MATTSON, *Rosenstiel School of Marine and Atmospheric Sciences, University of Miami, Florida*, and HARRY B. MARK, JR., *Department of Chemistry, University of Cincinnati, Ohio*

248 pages, illustrated. 1971

PATRICK Treatise on Adhesion and Adhesives

edited by ROBERT L. PATRICN, *Alpha Research and Development, Inc., Blue Island, Illinois*

Vol. 1 Theory
496 pages, illustrated. 1967
Vol. 2 Materials
568 pages, illustrated. 1969
Vol. 3 Special Topics
264 pages, illustrated. 1973

SCHICK Nonionic Surfactants

(Surfactant Science Series, Volume 1)

edited by MARTIN J. SCHICK, *Central Research Laboratories, Interchemical Corporation, Clifton, New Jersey*

1,120 pages, illustrated. 1967

SHINODA Solvent Properties of Surfactant Solutions

(Surfactant Science Series. Volume 2)
edited by KOZO SHINODA, *Department of Chemistry, Yokohama National University, Japan*
376 pages, illustrated. 1967

SLADE and JENKINS
Thermal Analysis

(Techniques and Methods of Polymer Evaluation Series, Volume 1)
edited by PHILIP E. SLADE, JR., *Monsanto Company, Pensacola, Florida,* and LLOYD T. JENKINS, *Chemstrand Research Center, Durham, North Carolina*
264 pages, illustrated. 1966

SLADE and JENKINS
Thermal Characterization Techniques

(Techniques and Methods of Polymer Evaluation Series, Volume 2)
edited by PHILIP E. SLADE, JR., *Monsanto Company, Pensacola, Florida* and LLOYD T. JENKINS, *Chemstrand Research Center, Durham, North Carolina*
384 pages, illustrated. 1970

STEVENS Characterization and Analysis of Polymers by Gas Chromatography

(Techniques and Methods of Polymer Evaluation Series, Volume 3)
by MALCOLM P. STEVENS, *American University of Beirut, Lebanon*
216 pages, illustrated. 1969

SWISHER Surfactant Biodegradation

(Surfactant Science Series, Volume 3)
by R. D. SWISHER, *Monsanto Company, St. Louis, Missouri*
520 pages, illustrated. 1970

WALTON Radome Engineering Handbook: *Design and Principles*

(Ceramics and Glass: Science and Technology Series, Volume 1)
edited by JESSE D. WALTON, JR., *Georgia Institute of Technology, Atlanta*
616 pages, illustrated. 1970

JOURNALS OF INTEREST

BIOMATERIALS, MEDICAL DEVICES, AND ARTIFICIAL ORGANS
An International Journal

editor: T. F. YEN, *University of Southern California, Los Angeles*

The aim of this new international journal is to bridge the gap between the theoretical aspects and practical applications of artificial organs and other medical devices, and implantation materials. The basic principles responsible for the success of artificial organs are stressed in order to encourage new research in this field.

4 issues per volume

JOURNAL OF MACROMOLECULAR SCIENCE—Chemistry

editor: GEORGE E. HAM, *White Plains, New York*

This international journal provides scientists with a cross-section of the outstanding contributions from laboratories around the world—published four and one-half months from publisher's receipt of last manuscript. The fields covered include anionic, cationic, and free-radical addition polymerization and copolymerization, the manifold forms of condensation polymerization, polymer reactions, molecular weight studies, temperature-dependent properties, rheology, effects of radiation of all forms, polymer degradation, and many others.

8 issues per volume

JOURNAL OF MACROMOLECULAR SCIENCE—*Physics*

editor: PHILLIP H. GEIL, *Case Western Reserve University*

A periodical devoted to the publication of significant fundamental contributions concerning the physics of macromolecular solids and liquids. Papers deal with research in transition mechanisms and structure property relationships, the physics of polymer solutions and melts, glassy and rubbery amorphous solids, and individual polymer molecules and natural polymers, as well as all the areas generally contained in polymer state physics.

4 issues per volume

JOURNAL OF MACROMOLECULAR SCIENCE—*Reviews in Macromolecular Chemistry*

editors: GEORGE B. BUTLER, *University of Florida, Gainesville,* and KENNETH F. O'DRISCOLL, *State University of New York, Buffalo,* and MITCHELL SHEN, *University of California, Berkeley*

Topics in this journal are reviews of certain recent chronological periods, and also have the advantage of reflecting the authors' knowledge, interpretation, and concise summary of the state of knowledge in the given area. Because of the nature of the journal, and the short time between completion of a manuscript and its publication, reviews of this nature more closely approximate a current review than could otherwise be accomplished.

2 issues per volume

POLYMER-PLASTICS TECHNOLOGY AND ENGINEERING

editor: LOUIS NATURMAN, *Stamford, Connecticut*

This journal reflects the increasing importance that polymer applications, processing developments, and mass production of new polymer products will have in the coming years. The emphasis of the articles that comprise the journal will also consider plastics technology and engineering as an important new feature of this publication.

2 issues per volume

Examination On-Approval Policy

Our policy allows instructors to examine a particular book for a period of two months without charge. In the event that the book is definitely adopted for a course as a class text, the instructor may retain the copy as *his desk copy,* provided he advises us of the adoption and the number of students enrolled in the class. If, however, the book will not be used as a class text, the instructor may return it or send us his remittance, less the educational discount.

JOURNAL SUBSCRIPTION INFORMATION

Subscriptions are entered on a calendar year basis. When a subscription is entered, it entitles the subscriber to all issues in the particular volume.

All journal subscriptions are processed after payment has been received. Only prepaid orders will receive service.

Cancellations requested for other than publisher's error will be accepted only prior to the publication of the first issue of each current volume, and will be subject to a handling charge.

Please add **foreign postage** for delivery to all countries outside the U.S. and Canada.

Air mail postage is available upon request.

Indexes to journals are bound into the last issue of each volume with the exception of review journals which are not indexed.

Back volume information and prices are available upon request.

A complimentary copy of any journal is available upon request.

- -

Date_____

MARCEL DEKKER, INC.
95 Madison Avenue
New York, New York 10016

Please send me a complimentary copy of the following journal(s):

_____

_____

_____

Name_____

Position_____

Company_____

Address_____

City_____ State_____ Zip_____